PRACTICE - ASSESS - DIAGNOSE

180 Days of GEOGRAPHY for Fifth Grade

Author
Kristin Kemp, M.A.Ed.

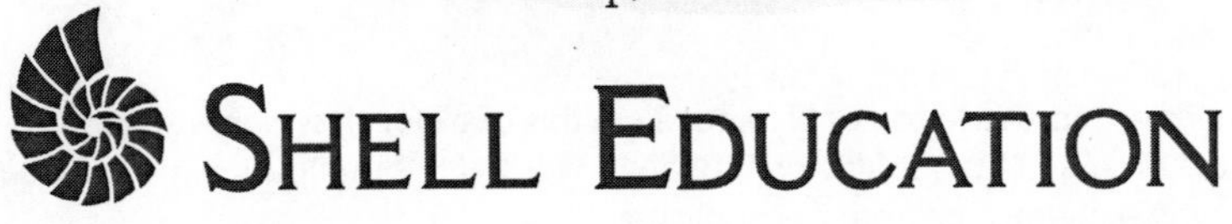

Series Consultant

Nicholas Baker, Ed.D.
Supervisor of Curriculum and Instruction

Colonial School District, DE

Publishing Credits

Corinne Burton, M.A.Ed., *Publisher*
Conni Medina, M.A.Ed., *Managing Editor*
Emily R. Smith, M.A.Ed., *Content Director*
Veronique Bos, *Creative Director*
Shaun N. Bernadou, *Art Director*
Lynette Ordoñez, *Editor*
Kevin Pham, *Graphic Designer*
Stephanie Bernard, *Associate Editor*

Image Credits

p.27 JHMimaging/Shutterstock; p.32 John Kropewnicki/Shutterstock; p.47 Gabes1976/iStock; p.52 Meinzahn/iStock; p.53 (left) Library of Congress [LC-D4271-353]; p.53 (right) Amlan Mathur/iStock; p.77 Kokkai/iStock; p.81 (left) Corey Leopold; p.81 (right) NASA; p.82 Simon Bradfield/iStock; p.118 Library of Congress [LC-USZ62-104098]; p.127 Library of Congress [LC-USZ62-93985]; p.137 Brytta/iStock; p.142 Library of Congress [LC-DIG-ggbain-20335]; p.187 Cn0ra/iStock; all other images from iStock and/or Shutterstock.

Standards

For information on how this resource meets national and other state standards, see pages 10–14. You may also review this information by visiting our website at www.teachercreatedmaterials.com/administrators/correlations/ and following the on-screen directions.

Shell Education
A division of Teacher Created Materials
5301 Oceanus Drive
Huntington Beach, CA 92649-1030
www.tcmpub.com/shell-education

ISBN 978-1-4258-3306-0

TABLE OF CONTENTS

INTRODUCTION

With today's geographic technology, the world seems smaller than ever. Satellites can accurately measure the distance between any two points on the planet and give detailed instructions about how to get there in real time. This may lead some people to wonder why we still study geography.

While technology is helpful, it isn't always accurate. We may need to find detours around construction, use a trail map, outsmart our technology, and even be the creators of the next navigational technology.

But geography is also the study of cultures and how people interact with the physical world. People change the environment, and the environment affects how people live. People divide the land for a variety of reasons. Yet no matter how it is divided or why, people are at the heart of these decisions. To be responsible and civically engaged, students must learn to think in geographical terms.

The Need for Practice

To be successful in geography, students must understand how the physical world affects humanity. They must not only master map skills but also learn how to look at the world through a geographical lens. Through repeated practice, students will learn how a variety of factors affect the world in which they live.

Understanding Assessment

In addition to providing opportunities for frequent practice, teachers must be able to assess students' geographical understandings. This allows teachers to adequately address students' misconceptions, build on their current understandings, and challenge them appropriately. Assessment is a long-term process that involves careful analysis of student responses from a discussion, project, practice sheet, or test. The data gathered from assessments should be used to inform instruction: slow down, speed up, or reteach. This type of assessment is called *formative assessment.*

HOW TO USE THIS BOOK

Weekly Structure

The first two weeks of the book focus on map skills. By introducing these skills early in the year, students will have a strong foundation on which to build throughout the year. The last two weeks allow students to practice naming states and capitals. Each of the remaining 30 weeks will follow a regular weekly structure.

Each week, students will study a grade-level geography topic and a location in the world. Locations may be a town, a state, a region, a continent, or the whole world.

Days 1 and 2 of each week focus on map skills. Days 3 and 4 allow students to apply information and data to what they have learned. Day 5 helps students connect what they have learned to themselves.

Day 1—Reading Maps: Students will study a grade-appropriate map and answer questions about it.

Day 2—Creating Maps: Students will create maps or add to an existing map.

Day 3—Read About It: Students will read a text related to the topic or location for the week and answer text-dependent or photo-dependent questions about it.

Day 4—Think About It: Students will analyze a chart, diagram, or other graphic related to the topic or location for the week and answer questions about it.

Day 5—Geography and Me: Students will do an activity to connect what they learned to themselves.

Five Themes of Geography

Good geography teaching encompasses all five themes of geography: location, place, human-environment interaction, movement, and region. Location refers to physical and absolute and relative locations or a specific point or place. The place theme refers to the human characteristics of a place. Human-environment interaction describes how humans affect their surroundings and how the environment affects the people who live there. Movement describes how and why people, goods, and ideas move between different places. The region theme examines how places are grouped into different regions. Regions can be divided based on a variety of factors, including physical characteristics, cultures, weather, political factors, and many others.

HOW TO USE THIS BOOK *(cont.)*

Weekly Themes

The following chart shows the topics, locations, and themes of geography that are covered during each week of instruction.

	Topic	Location	Geography Themes
1	—Map Skills Only—		Location
2			Location
3	Recycling	Mexico (N. America)	Human-Environment Interaction
4	Pony Express	United States (N. America)	Movement
5	Politics	United States (N. America)	Place, Region
6	Appalachian Region	Canada (N. America)	Place, Region
7	Tornados	Tornado Alley (N. America)	Human-Environment Interaction, Region
8	Trade	India (Asia)	Place, Movement
9	Religion	Israel (Asia)	Place
10	Housing	Japan (Asia)	Place
11	Population	China (Asia)	Place, Region
12	Vegetation and Climate	Southeast Asia	Place, Region
13	Transportation	Australia	Movement
14	Uluru	Northern Territory (Australia)	Location, Place
15	Climate	Australia	Location, Human-Environment Interaction
16	Population	Australia	Movement
17	Absolute and Relative Locations	Australia	Location
18	Natural Resources	Venezuela (S. America)	Human-Environment Interaction, Movement

HOW TO USE THIS BOOK *(cont.)*

	Topic	Location	Geography Themes
19	Regions	Peru (S. America)	Region
20	Rivers	Amazon River (S. America)	Place
21	Landforms	Argentina (S. America)	Location, Place, Region
22	Transportation	South America	Movement
23	Natural Resources	South Africa	Place, Human-Environment Interaction
24	Landmarks	Egypt (Africa)	Place
25	Language	Zimbabwe (Africa)	Place, Region
26	Imperialism	Africa	Location, Region
27	Climate	Africa	Place, Human-Environment Interaction, Region
28	Coastal Erosion	France (Europe)	Place, Human-Environment Interaction
29	Languages	Baltic States (Europe)	Location, Place, Region
30	Renewable Resources	Scandinavian Countries (Europe)	Human-Environment Interaction, Region
31	Imports and Exports	United Kingdom (Europe)	Movement
32	Language	Greece (Europe)	Place
33	Immigration	World	Movement
34	Latitude and Longitude	World	Location, Movement
35	States	United States	Location
36	State Capitals	United States	Location

HOW TO USE THIS BOOK *(cont.)*

Using the Practice Pages

The activity pages provide practice and assessment opportunities for each day of the school year. Teachers may wish to prepare packets of weekly practice pages for the classroom or for homework.

As outlined on page 4, each week examines one location and one geography topic.

The first two days focus on map skills. On Day 1, students will study a map and answer questions about it. On Day 2, they will add to or create a map.

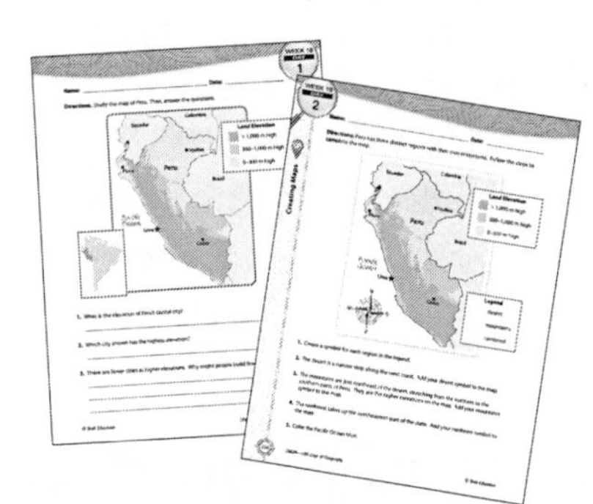

Days 3 and 4 allow students to apply information and data from texts, charts, graphs, and other sources to the location being studied.

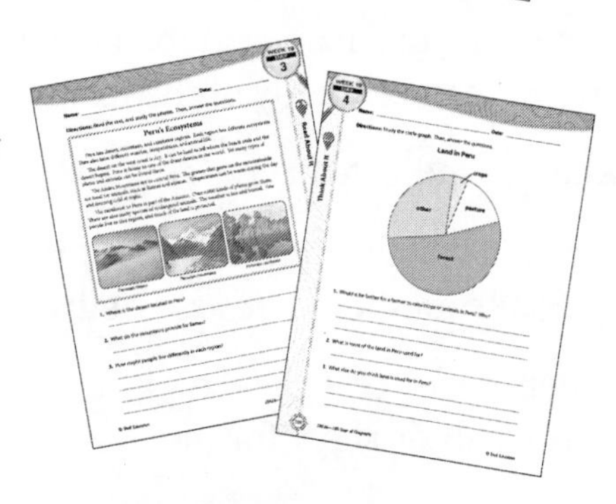

On Day 5, students will apply what they learned to themselves.

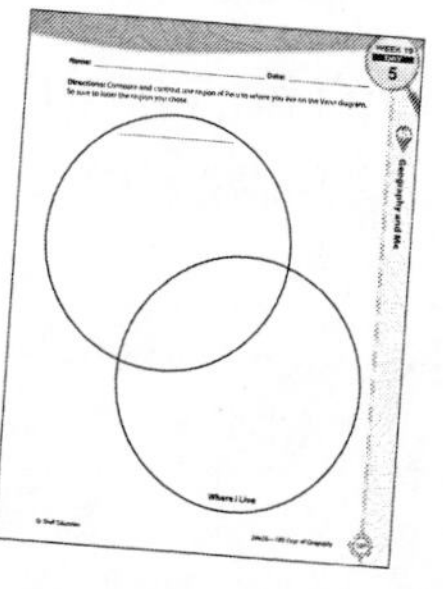

Using the Resources

Rubrics for the types of days (map skills, applying information and data, and making connections) can be found on pages 210–212 and in the Digital Resources. Use the rubrics to assess students' work. Be sure to share these rubrics with students often so that they know what is expected of them.

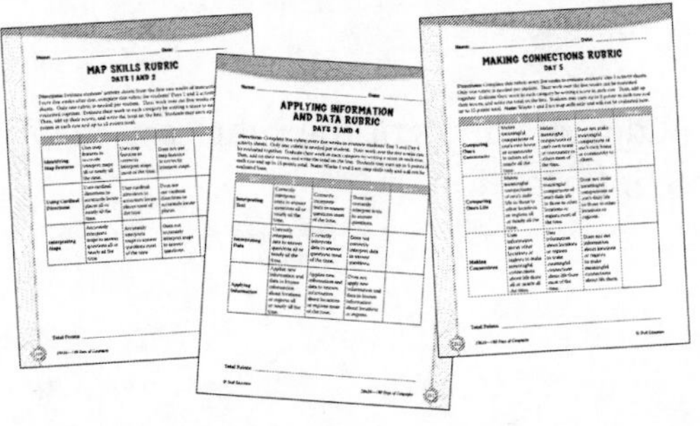

HOW TO USE THIS BOOK *(cont.)*

Diagnostic Assessment

Teachers can use the practice pages as diagnostic assessments. The data analysis tools included with the book enable teachers or parents to quickly score students' work and monitor their progress. Teachers and parents can quickly see which skills students may need to target further to develop proficiency.

Students will learn map skills, how to apply text and data to what they have learned, and how to relate what they learned to themselves. Teachers can assess students' learning in each area using the rubrics on pages 210–212. Then, record their scores on the Practice Page Item Analysis sheets on pages 213–215. These charts are also provided in the Digital Resources as PDFs, Microsoft Word® files, and Microsoft Excel® files (see page 216 for more information). Teachers can input data into the electronic files directly on the computer, or they can print the pages.

To Complete the Practice Page Item Analyses:

- Write or type students' names in the far-left column. Depending on the number of students, more than one copy of the forms may be needed.
    - The skills are indicated across the tops of the pages.
    - The weeks in which students should be assessed are indicated in the first rows of the charts. Students should be assessed at the ends of those weeks.
- Review students' work for the days indicated in the chart. For example, if using the Making Connections Analysis sheet for the first time, review students' work from Day 5 for all five weeks.
- Add the scores for each student. Place that sum in the far right column. Record the class average in the last row. Use these scores as benchmarks to determine how students are performing.

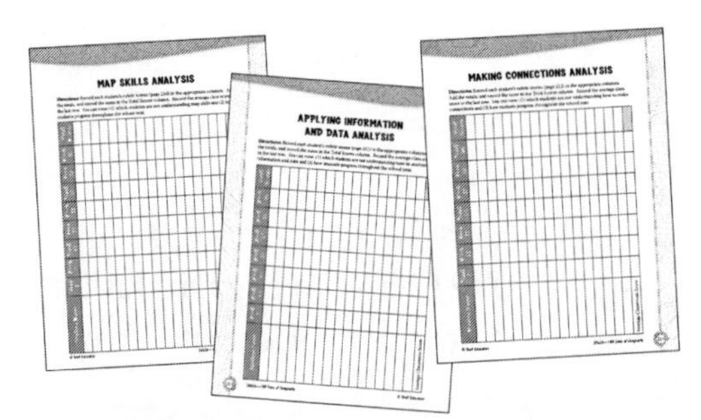

Digital Resources

The Digital Resources contain digital copies of the rubrics, item analysis sheets, and standards charts. See page 216 for more information.

HOW TO USE THIS BOOK *(cont.)*

Using the Results to Differentiate Instruction

Once results are gathered and analyzed, teachers can use them to inform the way they differentiate instruction. The data can help determine which geography skills are the most difficult for students and which students need additional instructional support and continued practice.

Whole-Class Support

The results of the diagnostic analysis may show that the entire class is struggling with certain geography skills. If these concepts have been taught in the past, this indicates that further instruction or reteaching is necessary. If these concepts have not been taught in the past, this data is a great preassessment and may demonstrate that students do not have a working knowledge of the concepts. Thus, careful planning for the length of the unit(s) or lesson(s) must be considered, and additional front-loading may be required.

Small-Group or Individual Support

The results of the diagnostic analysis may show that an individual student or a small group of students is struggling with certain geography skills. If these concepts have been taught in the past, this indicates that further instruction or reteaching is necessary. Consider pulling these students aside to instruct them further on the concepts while others are working independently. Students may also benefit from extra practice using games or computer-based resources.

Teachers can also use the results to help identify proficient individual students or groups of students who are ready for enrichment or above-grade-level instruction. These students may benefit from independent learning contracts or more challenging activities.

STANDARDS CORRELATIONS

Shell Education is committed to producing educational materials that are research and standards based. In this effort, we have correlated all our products to the academic standards of all 50 states, the District of Columbia, the Department of Defense Dependents Schools, and all Canadian provinces.

How to Find Standards Correlations

To print a customized correlation report of this product for your state, visit our website at **www.teachercreatedmaterials.com/administrators/correlations** and follow the on-screen directions. If you require assistance in printing correlation reports, please contact our Customer Service Department at 1-877-777-3450.

Purpose and Intent of Standards

The Every Student Succeeds Act (ESSA) mandates that all states adopt challenging academic standards that help students meet the goal of college and career readiness. While many states already adopted academic standards prior to ESSA, the act continues to hold states accountable for detailed and comprehensive standards. Standards are designed to focus instruction and guide adoption of curricula. Standards are statements that describe the criteria necessary for students to meet specific academic goals. They define the knowledge, skills, and content students should acquire at each level. Standards are also used to develop standardized tests to evaluate students' academic progress. Teachers are required to demonstrate how their lessons meet state standards. State standards are used in the development of our products, so educators can be assured they meet the academic requirements of each state.

The activities in this book are aligned to the National Geography Standards and the McREL standards. The chart on pages 11–12 lists the National Geography Standards used throughout this book. The chart on pages 13–14 correlates the specific McREL and National Geography Standards to each week. The standards charts are also in the Digital Resources (standards.pdf).

C3 Framework

This book also correlates to the College, Career, and Civic Life (C3) Framework published by the National Council for the Social Studies. By completing the activities in this book, students will learn to answer and develop strong questions (Dimension 1), critically think like a geographer (Dimension 2), and effectively choose and use geography resources (Dimension 3). Many activities also encourage students to take informed action within their communities (Dimension 4).

STANDARDS CORRELATIONS *(cont.)*

180 Days of Geography is designed to give students daily practice in geography through engaging activities. Students will learn map skills, how to apply information and data to their understandings of various locations and cultures, and how to apply what they learned to themselves.

Easy to Use and Standards Based

There are 18 National Geography Standards, which fall under six essential elements. Specific expectations are given for fourth grade, eighth grade, and twelfth grade. For this book, eighth grade expectations were used with the understanding that full mastery is not expected until that grade level.

Essential Elements	National Geography Standards
The World in Spatial Terms	**Standard 1:** How to use maps and other geographic representations, geospatial technologies, and spatial thinking to understand and communicate information
	Standard 2: How to use mental maps to organize information about people, places, and environments in a spatial context
	Standard 3: How to analyze the spatial organization of people, places, and environments on Earth's surface
Places and Regions	**Standard 4:** The physical and human characteristics of places
	Standard 5: People create regions to interpret Earth's complexity
	Standard 6: How culture and experience influence people's perceptions of places and regions
Physical Systems	**Standard 7:** The physical processes that shape the patterns of Earth's surface
	Standard 8: The characteristics and spatial distribution of ecosystems and biomes on Earth's surface

STANDARDS CORRELATIONS *(cont.)*

Essential Elements	National Geography Standards
Human Systems	**Standard 9:** The characteristics, distribution, and migration of human populations on Earth's surface
	Standard 10: The characteristics, distribution, and complexity of Earth's cultural mosaics
	Standard 11: The patterns and networks of economic interdependence on Earth's surface
	Standard 12: The process, patterns, and functions of human settlement
	Standard 13: How the forces of cooperation and conflict among people influence the division and control of Earth's surface
Environment and Society	**Standard 14:** How human actions modify the physical environment
	Standard 15: How physical systems affect human systems
	Standard 16: The changes that occur in the meaning, use, distribution, and importance of resources
The Uses of Geography	**Standard 17:** How to apply geography to interpret the past
	Standard 18: How to apply geography to interpret the present and plan for the future

–2012 National Council for Geographic Education

STANDARDS CORRELATIONS *(cont.)*

Easy to Use and Standards Based *(cont.)*

This chart lists the specific National Geography Standards and McREL standards that are covered each week.

Wk.	NGS	McREL Standards
1	Standards 1 and 3	Knows the basic elements of maps and globes. Uses map grids to plot absolute location.
2	Standards 1 and 3	Knows the approximate location of major continents, mountain ranges, and bodies of water on Earth.
3	Standard 14	Knows advantages and disadvantages of recycling and reusing different types of materials.
4	Standard 11	Knows how transportation and communication have changed and how they have affected trade and economic activities.
5	Standard 13	Knows the similarities and differences in characteristics of culture in different regions.
6	Standard 4	Understands how physical processes help to shape features and patterns on Earth's surface.
7	Standard 15	Knows natural hazards that occur in the physical environment.
8	Standard 11	Knows how regions are linked economically and how trade affects the way people earn their living in each region.
9	Standard 10	Understands how different people living in the same region maintain different ways of life.
10	Standard 12	Knows the characteristics of a variety of regions.
11	Standard 9	Knows the spatial distribution of population.
12	Standard 5	Knows the characteristics of a variety of regions.
13	Standard 11	Knows how transportation and communication have changed and how they have affected trade and economic activities.
14	Standards 7 and 10	Knows the physical components of Earth's atmosphere, lithosphere, hydrosphere, and biosphere. Knows the similarities and differences in characteristics of culture in different regions.
15	Standard 7	Knows how Earth's position relative to the Sun affects events and conditions on Earth.
16	Standards 9 and 17	Understands the characteristics of populations at a variety of scales.

STANDARDS CORRELATIONS *(cont.)*

Wk.	NGS	McREL Standards
17	Standard 1	Knows the basic elements of maps and globes. Uses map grids to plot absolute location.
18	Standard 16	Knows economic activities that use natural resources in the local region, state, and nation and the importance of the activities to these areas.
19	Standard 5	Knows the characteristics of a variety of regions.
20	Standard 7	Knows the physical components of Earth's atmosphere, lithosphere, hydrosphere, and biosphere.
21	Standard 5	Knows the characteristics of a variety of regions.
22	Standard 11	Knows how transportation and communication have changed and how they have affected trade and economic activities.
23	Standards 11 and 16	Knows the relationships between economic activities and resources.
24	Standard 10	Knows the similarities and differences in characteristics of culture in different regions.
25	Standard 10	Knows the similarities and differences in characteristics of culture in different regions.
26	Standard 13	Knows how and why people compete for control of Earth's surface.
27	Standard 8	Knows the physical components of Earth's atmosphere, lithosphere, hydrosphere, and biosphere.
28	Standards 7 and 14	Knows the ways people alter the physical environment.
29	Standard 10	Knows the similarities and differences in characteristics of culture in different regions.
30	Standard 16	Knows the characteristics, location, and use of renewable resources, flow resources, and nonrenewable resources.
31	Standard 11	Knows the various ways in which people satisfy their basic needs and wants through the production of goods and services in different regions of the world.
32	Standard 10	Knows the similarities and differences in characteristics of culture in different regions.
33	Standard 9	Knows the causes and effects of human migration.
34	Standard 1	Uses map grids to plot absolute location.
35	Standard 2	Knows major physical and human features of places as they are represented on maps and globes.
36	Standard 2	Knows the location of major cities in North America.

Name: Caroline Bogan **Date:** 10/31/21

Directions: Use the map to answer the questions.

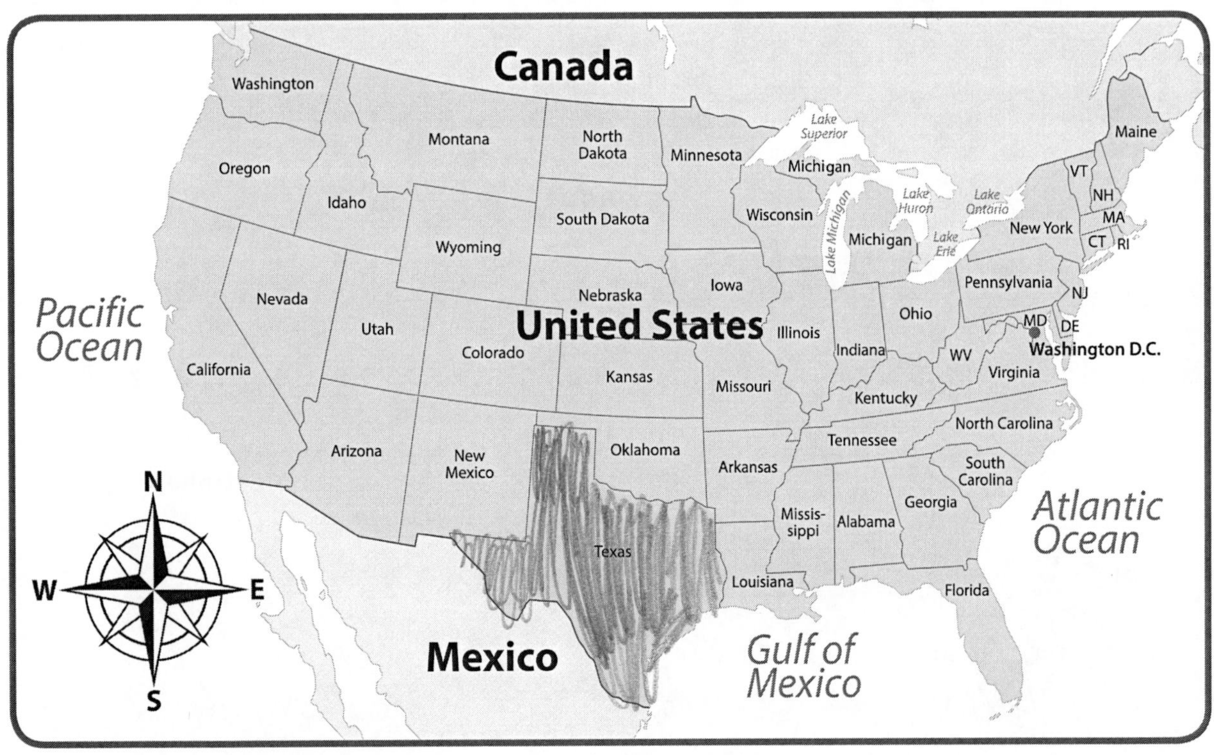

1. Shade a state you would like to visit.

2. What is north of that state?

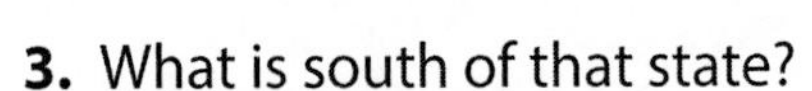

Oklahoma

3. What is south of that state?

Mexico

4. What is west of that state?

New Mexico

5. What is east of that state?

Oklahoma + Arkansas + Louisiana

6. Describe the size the state you chose.

The second largest state

Name: ______________________________ Date: ______________

Directions: Study the map of the world. Then, answer the questions.

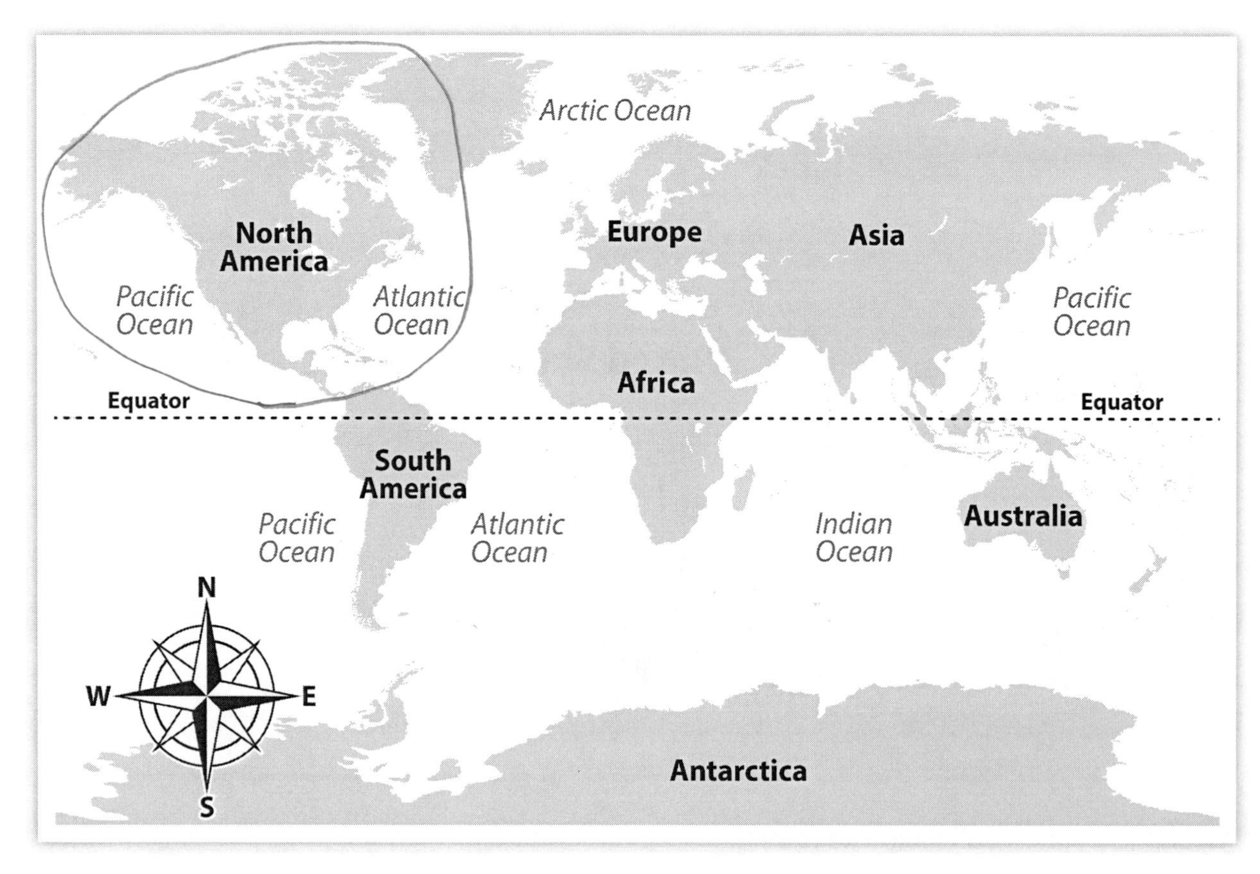

1. Circle the continent where you live.

2. Describe your continent's position. What other continents or bodies of water is it near? Use the compass rose to help you.

____

____

____

3. Trace the equator in blue.

4. Name two continents that are on the equator.

____

5. Name a continent that is entirely south of the equator.

____

Name: ______________________________ Date: ______________

Directions: Follow the steps to complete the map.

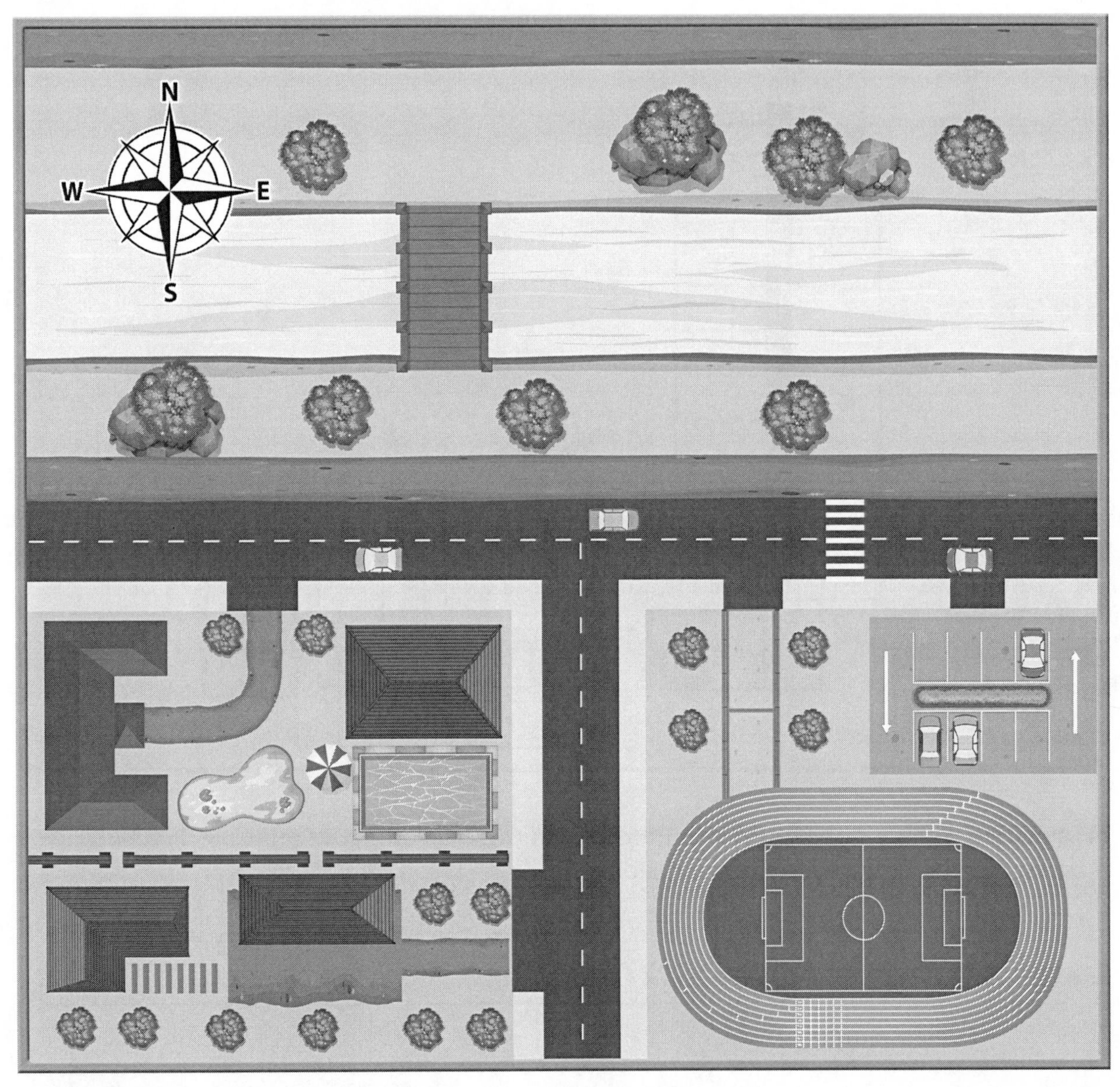

1. Add a boat in the river west of the bridge.
2. Draw a path someone might take from the stadium to the north side of the river.
3. Draw a fence around the house in the southwest corner of the map.
4. Draw a person in the northeast corner of the map.
5. Name each street.

Name: ______________________________ **Date:** ________________

Directions: Study the hotel fire escape map. Then, answer the questions.

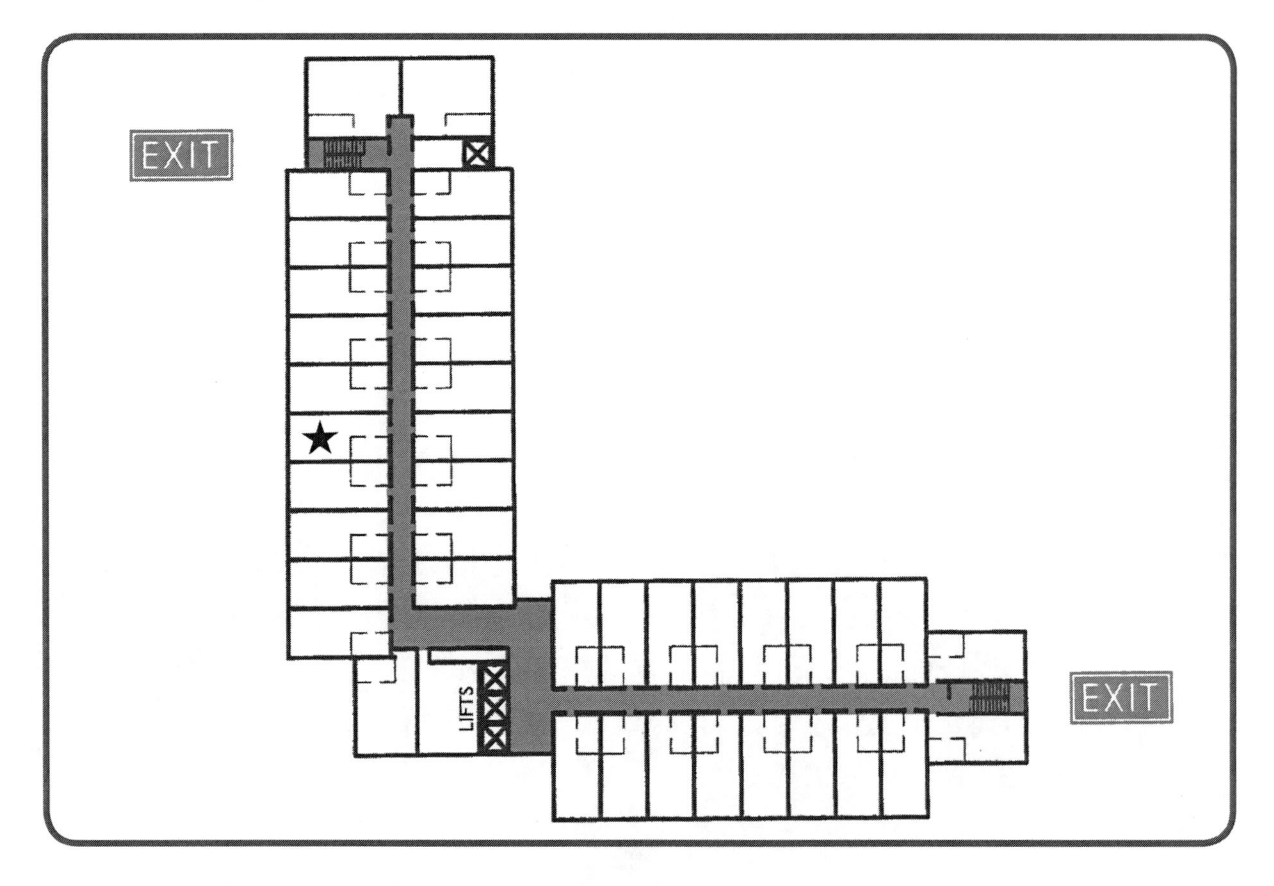

1. Imagine you are in the room with the star, and the fire alarm goes off. Give directions to describe the best way to exit the building.

____

____

____

____

2. Think about your classroom. Give directions to explain how to walk to a safe place outside.

____

____

____

____

____

Name: ______________________ Date: ______________

Directions: Draw the shapes on the grid.

	A	B	C	D
1				
2				
3				
4				

1. Draw a star in C1.
2. Draw a triangle in A3.
3. Draw a circle in D2.
4. Draw a diamond in B4.
5. Draw a heart in C3.
6. Draw a square in B1.

Name: ______________________________ Date: ____________________

Directions: Study the map of the world. Then, answer the questions.

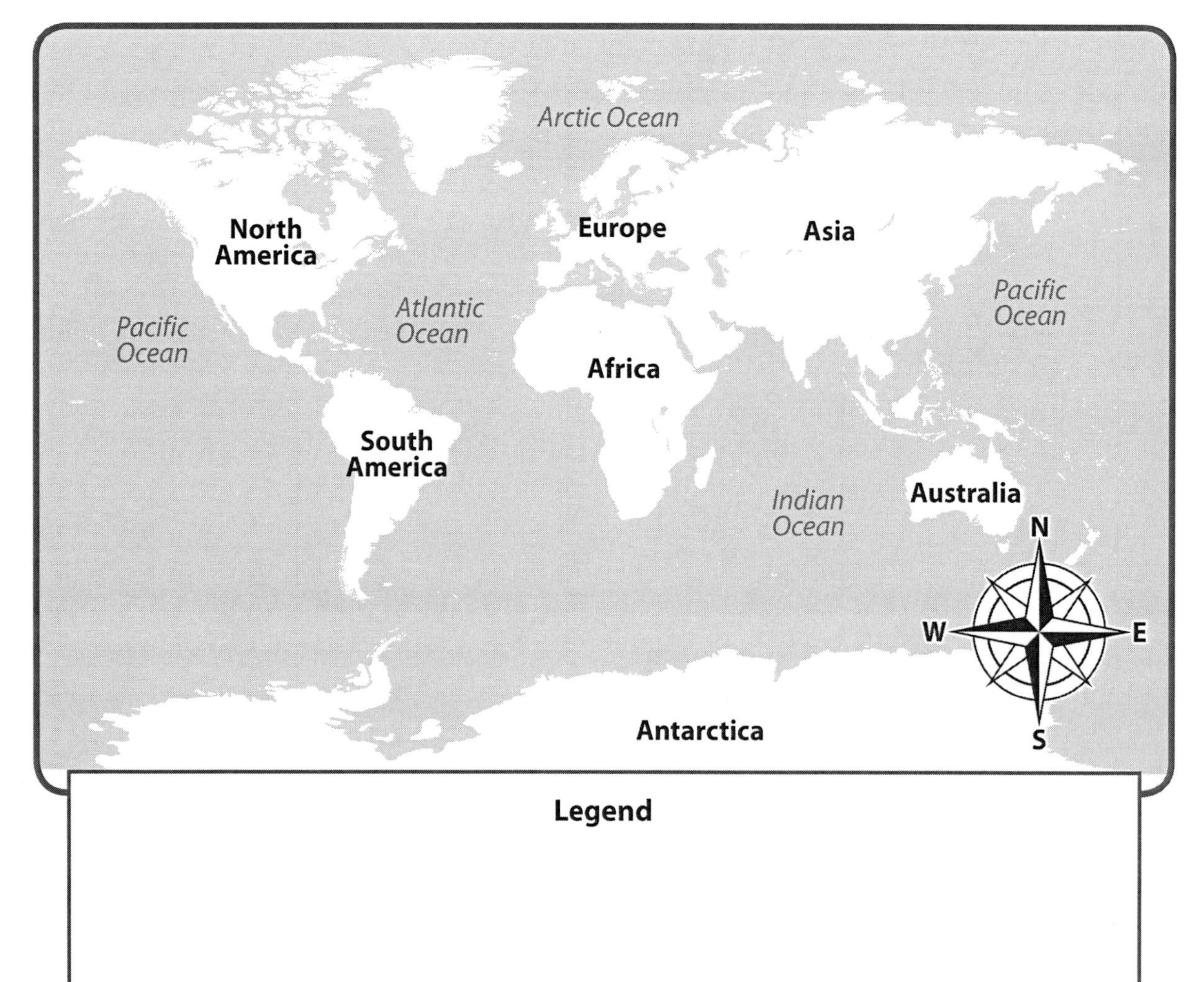

1. Which two continents are closest to Europe?

____

2. Which ocean is farthest north?

____

3. Name two oceans that border Asia.

____

4. Shade each continent a different color.

5. Create a legend to show which continent each color represents.

Name: ____________________ Date: ____________

Directions: Study this elevation map of Africa. Then, answer the questions.

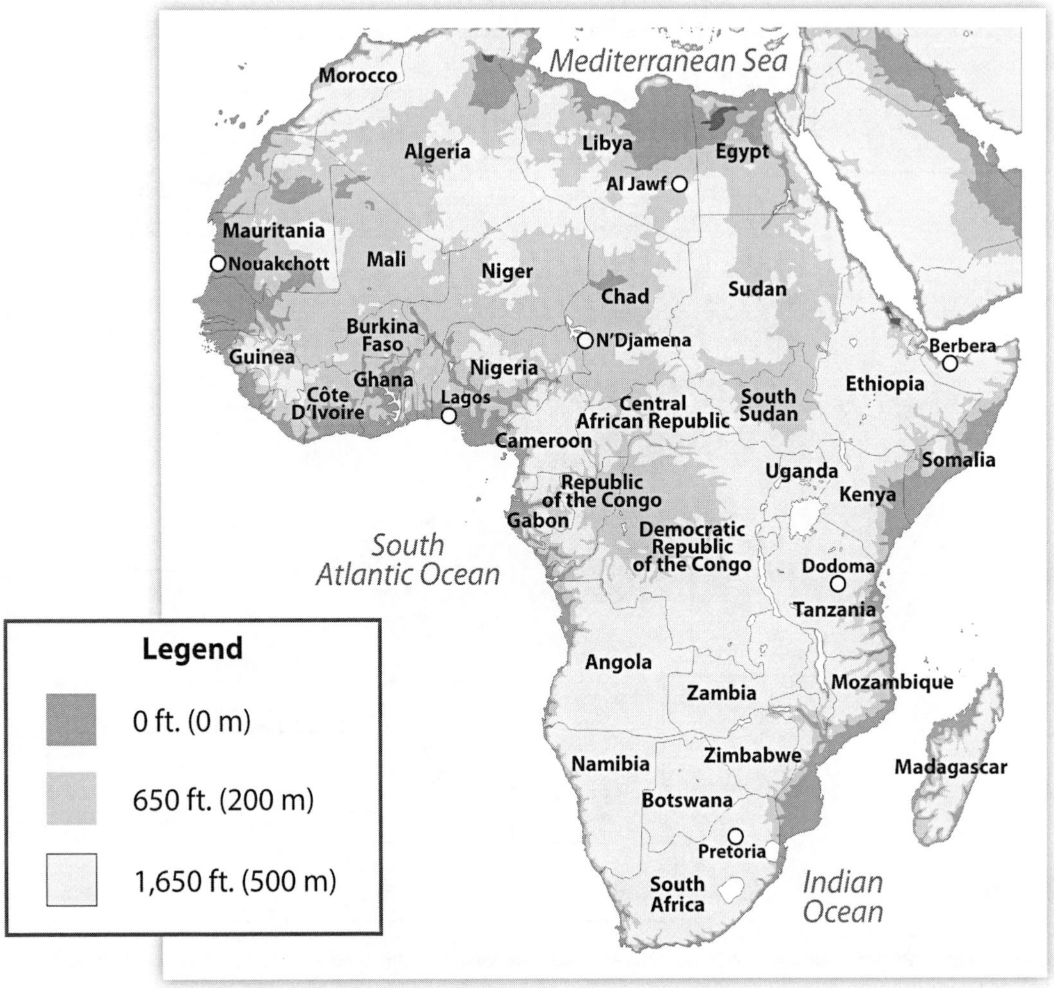

1. At what elevation is N'Djamena, Chad?

____

2. At what elevation is Lagos, Nigeria?

____

3. At what elevation is Pretoria, South Africa?

____

4. Describe the locations of areas that have an elevation of 0 ft. (0 m).

____

____

Map Skills

Name: ______________________________ **Date:** ________________

Directions: This map shows the states of Australia. Use the clues to label each state.

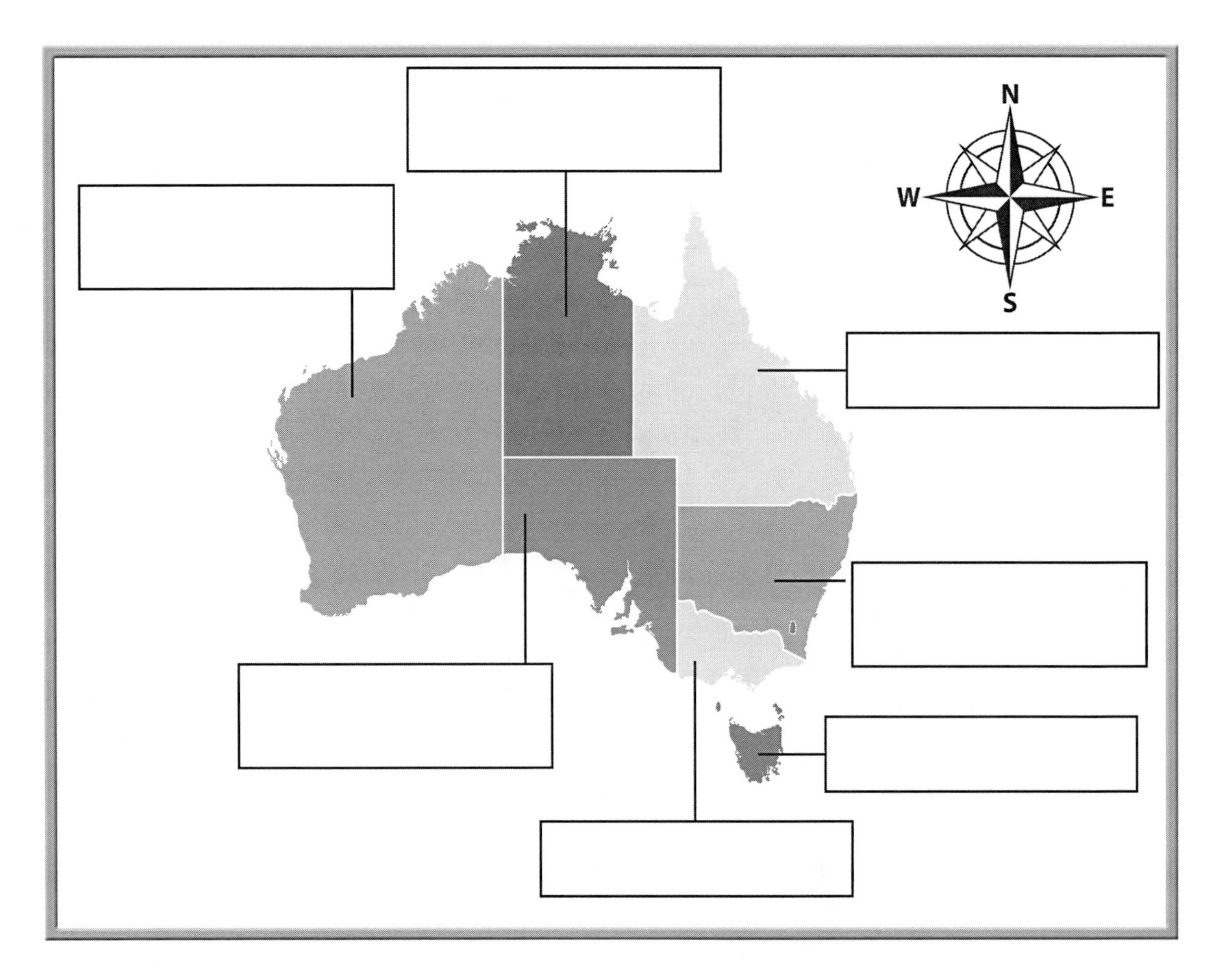

1. Tasmania is a small island off Australia's southeast coast.
2. Western Australia is a large, western state.
3. Victoria is a small state just north of Tasmania.
4. Northern Territory is in north central Australia.
5. Queensland is in northeastern Australia.
6. South Australia is in south central Australia.
7. New South Wales is north of Victoria.

Name: ____________________ Date: ____________

Directions: Draw a map of your community. Include places you go and landmarks. Use symbols to represent important places. Draw and label those symbols in the legend, too.

Legend

Map Skills

Name: ______________________________ **Date:** ________________

Directions: Some maps include lines of latitude and longitude. These lines help people find precise locations. Lines of latitude run east and west. Lines of longitude run north and south. Both are measured in degrees.

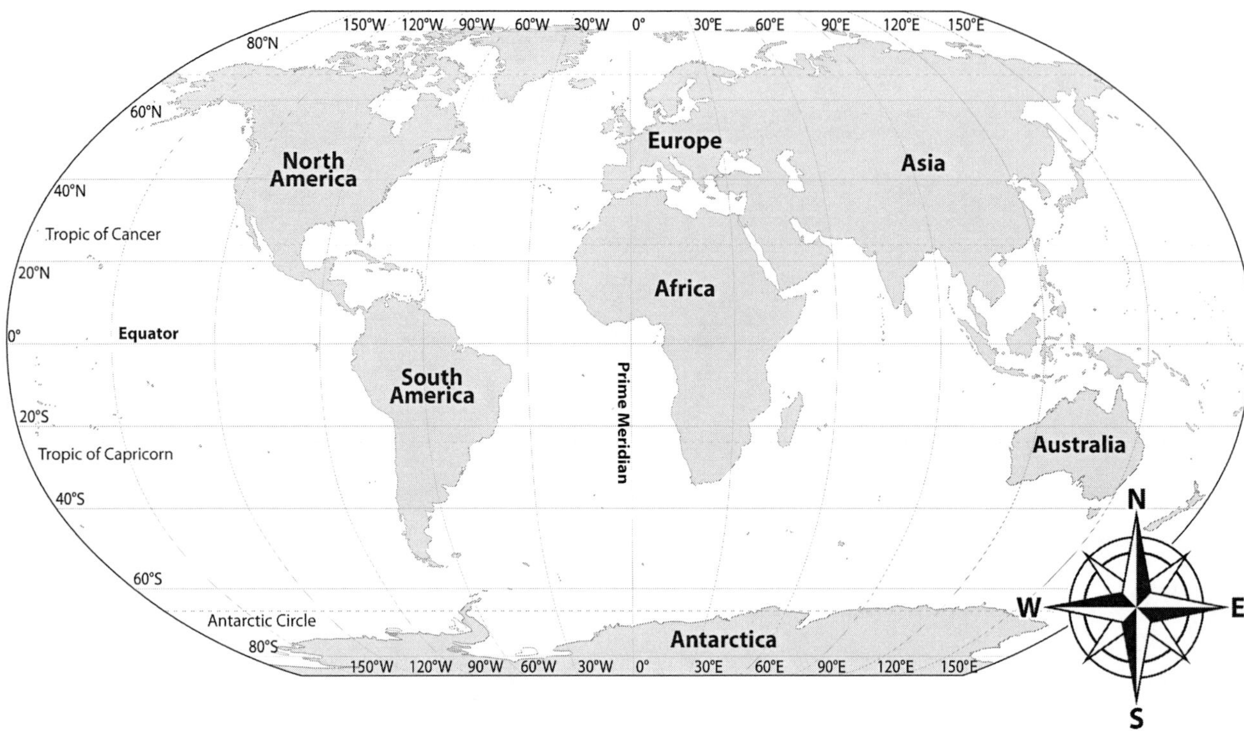

1. What is the latitude of the equator?

____

2. Which continent is located at 40°N and 90°E?

____

3. Which continent is located at 20°S and 60°W?

____

4. Which continent is located at 80°S?

____

5. Which continents does the Prime Meridian pass through?

____

Name: ______________________________ **Date:** ________________

Directions: Study the map of Mexico. Then, answer the questions.

1. Name at least one city on the Pacific coast.

____

2. Name at least one city on the coast of the Gulf of Mexico.

____

3. What is the capital of Mexico?

____

4. Name the three countries that border Mexico.

____

5. Name the southernmost city on this map.

____

Creating Maps

Name: ______________________________ **Date:** ________________

Directions: Use the clues to label the countries and bodies of water that border Mexico.

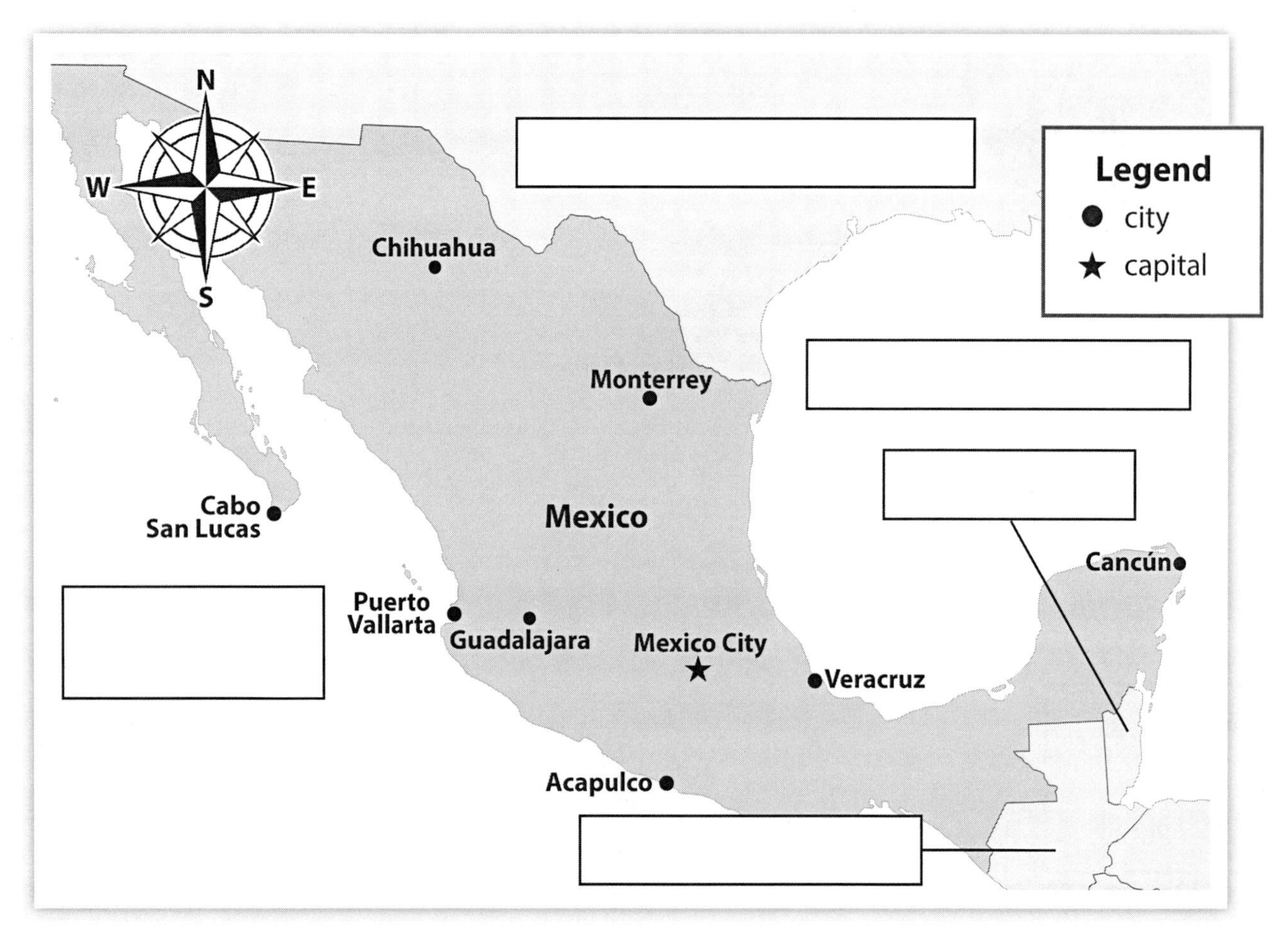

1. The United States borders Mexico to the north.

2. Belize borders Mexico to the southeast.

3. Guatemala also borders Mexico to the southeast. It is larger than Belize.

4. The Pacific Ocean is west of Mexico.

5. The Gulf of Mexico borders Mexico to the east.

Challenge: Study the city names on the map. Then, cover the city names with sticky notes. Label as many cities as you can before checking your answers.

Name: ______________________ Date: ______________

Directions: Read the text, and study the photo. Then, answer the questions.

Recycling for Food

Mexico City is a big city with a big problem—trash! A landfill opened in 1985 near the city. But by 2008, the landfill was full and needed to be closed. People kept bringing trash anyway. Three years later, the landfill finally closed. Now what would people do? They began illegally dumping their trash, burning it, or letting it pile up along the roads.

Recycling was not very popular in Mexico City, but leaders knew it could help solve the trash problem. They came up with a new idea. People could bring their bottles, cardboard, and glass to be recycled. In return, they would get free fruits and vegetables from local farmers. This answer had many benefits. It has allowed tons of trash to be recycled instead of taken to a trash dump. The people receive food they need. Farmers earn money from the government. Everyone wins!

1. What caused the problem with trash?

____

____

2. What was the leaders' plan to solve the trash problem?

____

____

3. How did this plan help people?

____

____

____

____

Name: ______________________ Date: ____________

Directions: The table shows how much trash people in different countries make each day. Study the table, and answer the questions.

Trash Around the World	
Country	**Kilograms of Trash per Person**
Australia	2.25
Brazil	1.0
China	1.0
Finland	2.1
India	0.3
Japan	1.65
Mexico	1.25
Turkey	1.75
United States	2.6

1. Which country produces the most trash per person?

____

2. Which country produces the least trash per person?

____

3. Write the countries in order from most trash to least trash per person.

____

____

____

____

4. How does this chart show the importance of recycling?

____

____

____

____

Name: ______________________ Date: ______________

Directions: Residents of Mexico City recycle to receive free fruits and vegetables. How could people in the United States be encouraged to recycle? List and illustrate two ways.

____

____

____

____

Name: ______________________________ **Date:** ______________

Directions: The Pony Express delivered mail by horseback from 1860 to 1861. Study the map showing its route. Then, answer the questions.

Pony Express

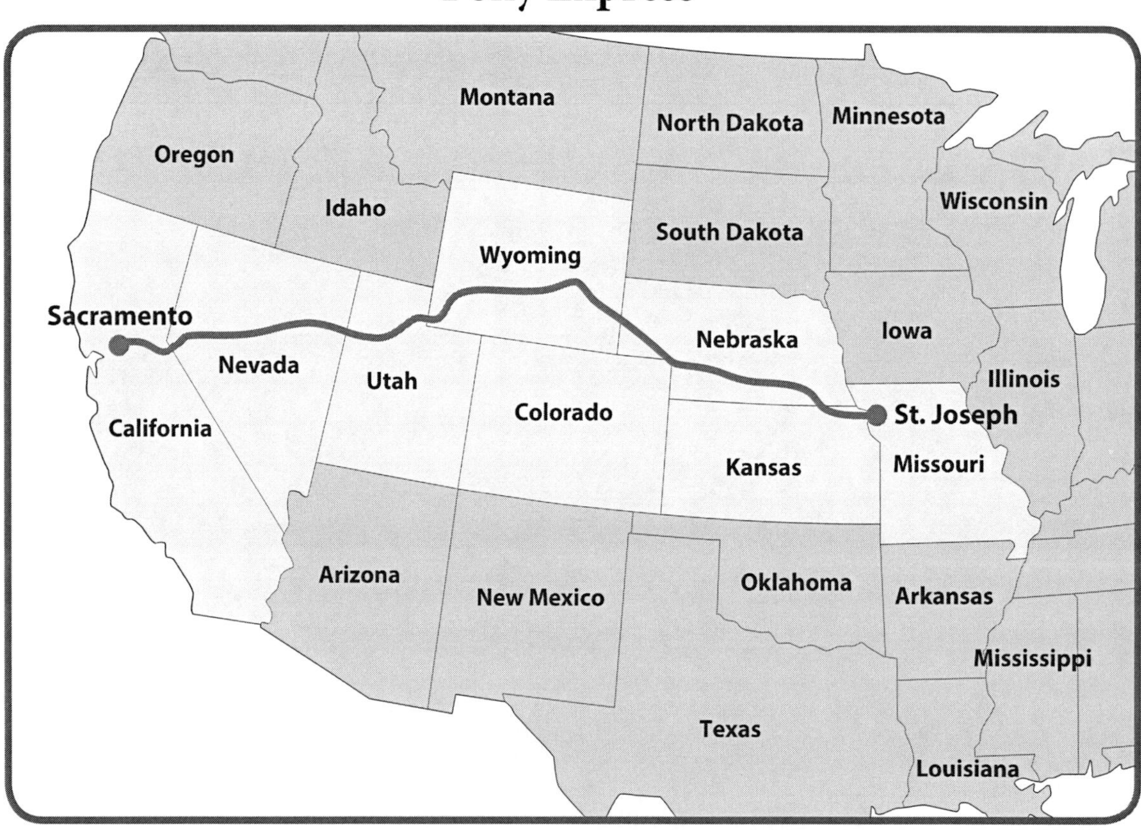

1. Where did the Pony Express begin and end?

____

2. What modern-day states did the Pony Express travel through?

____

____

____

3. Riders crossed the Rocky Mountains on this route. How do you think that affected their trips?

____

____

____

Name: ______________________ Date: ______________

Directions: Follow the steps to color the states that were part of the Pony Express's route.

Pony Express

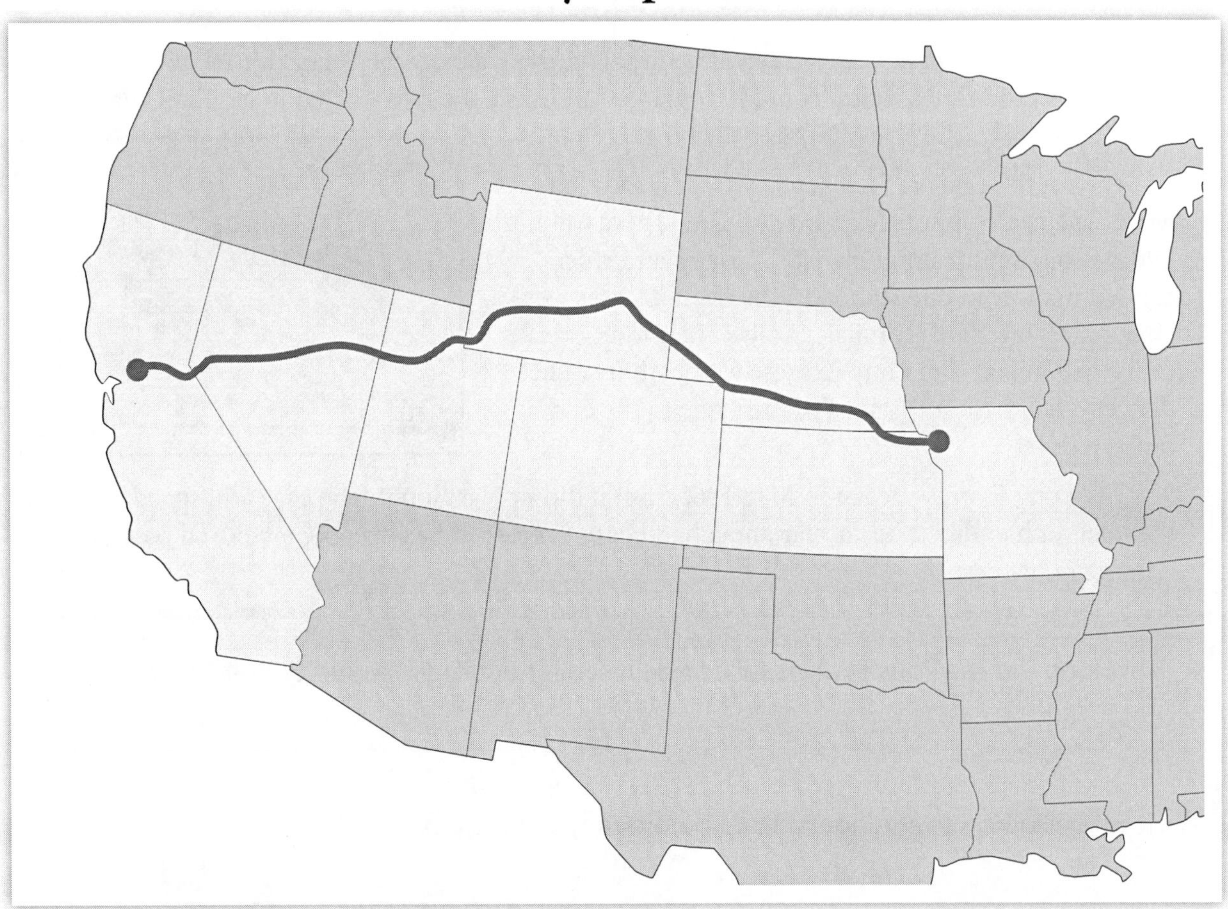

1. Color California green.
2. Color Colorado yellow.
3. Color Kansas orange.
4. Color Missouri red.
5. Color Nebraska blue.
6. Color Nevada purple.
7. Color Utah brown.
8. Color Wyoming pink.

Read About It

Name: ______________________ **Date:** ______________

Directions: Read the text, and study the photo. Then, answer the questions.

The Pony Express

People needed patience to receive their mail in the 1800s. A stagecoach took more than three weeks to bring a letter from Missouri to California. People wanted to get their mail faster than that. The Pony Express was born!

The route began in St. Joseph, MO, and traveled west. The last stop was Sacramento, CA. There were 190 stations set up along the way. Horses and riders carried mail from one station to the next. Horses were changed every 10 to 15 miles. A new rider took over every 100 miles. The Pony Express was fast! It could deliver a letter from the start of the route to the end in only 10 days.

The Pony Express began in May 1860. But it did not last long. The ease and speed of the telegraph ended it just a year and a half later. Even so, it has become a beloved part of American history.

1. How long did the Pony Express take to deliver a letter from Missouri to California?

____

2. How far did horses and riders travel before they got a break?

____

____

3. Why did the Pony Express only last a short time?

____

____

____

4. Why do you think the Pony Express is a beloved part of American history?

____

____

____

Name: ______________________________ Date: ______________

Directions: Study the photo of the Pony Express rider, and answer the questions.

Think About It

1. Why are there bags on the saddle?

____

____

2. Why do you think the statue was created?

____

____

____

3. What can you infer about the Pony Express from this photo?

____

____

____

Name: ______________________ Date: ______________

Directions: Complete the table comparing ways to communicate in the past and present. Make predictions about how people might communicate in the future.

Communication in the Past	Communication in the Present	Communication in the Future

Name: ______________________________ **Date:** ________________

Directions: In 2016, people in all 50 states voted for candidates in the presidential election. Most people either voted for the Democratic or the Republican candidates. This map shows the results of the election. Study the map, and answer the questions.

2016 Election Results

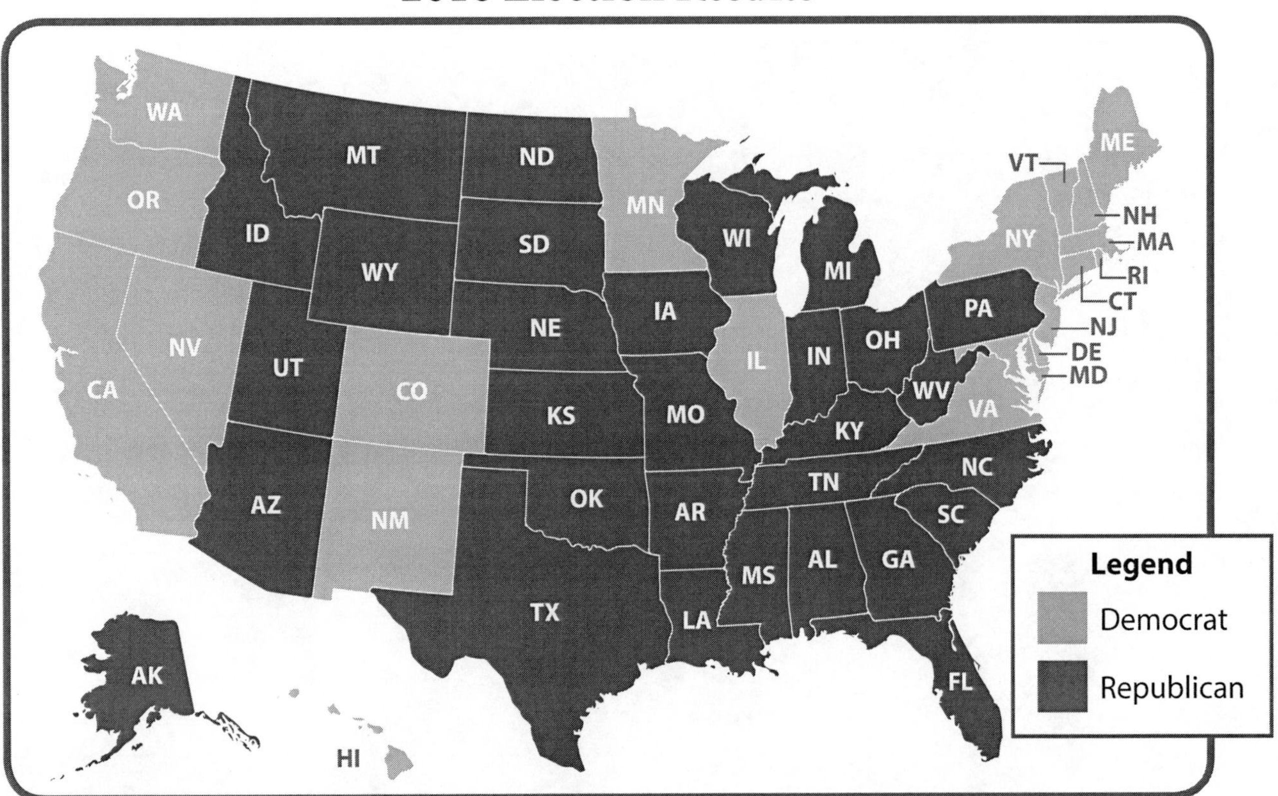

1. Name three states that voted for the Democratic candidate.

   ____

2. Name three states that voted for the Republican candidate.

   ____

3. Describe the locations of the Democratic states.

   ____

   ____

4. Describe the locations of the Republican states.

   ____

   ____

Name: ______________________ Date: ______________

Directions: Write the names of the states missing from the map.

2016 Election Results

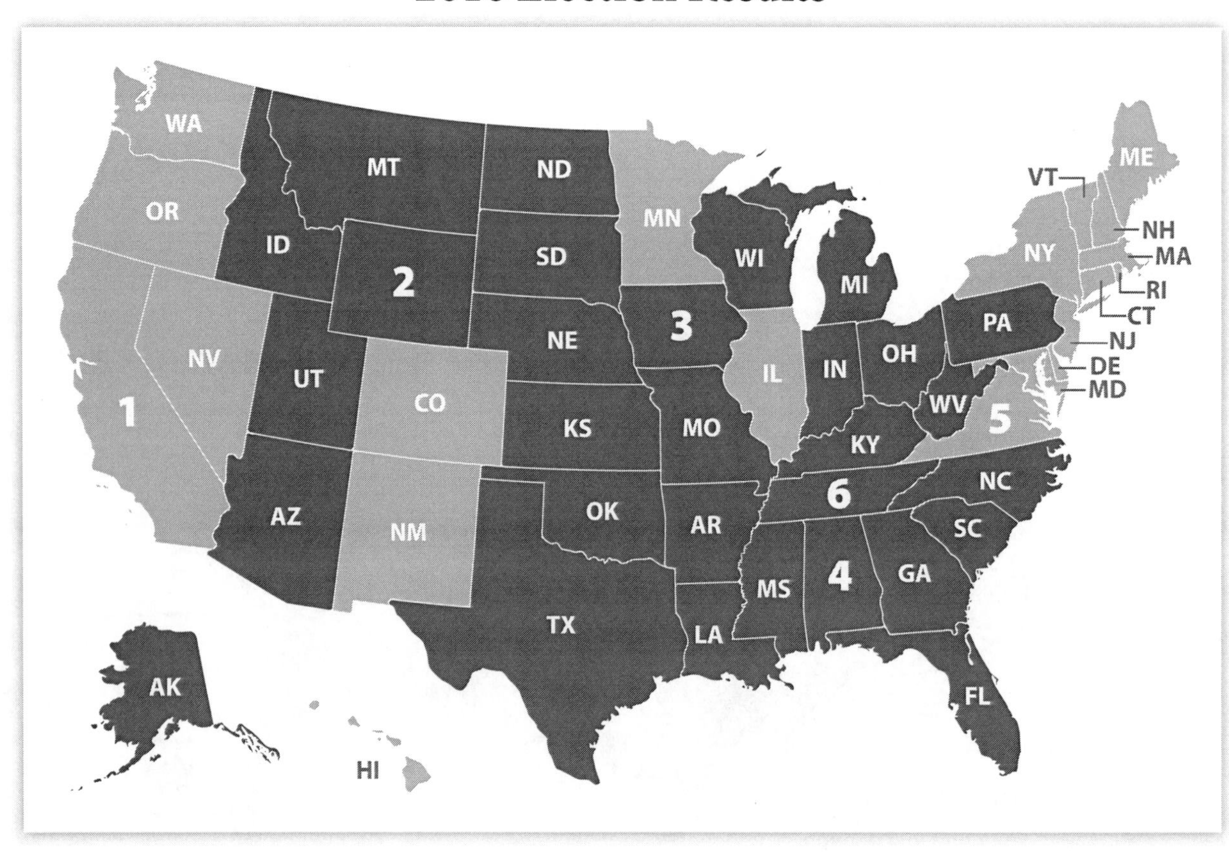

1. ______________________________

2. ______________________________

3. ______________________________

4. ______________________________

5. ______________________________

6. ______________________________

Challenge: Cover 10 more states with small slips of paper. Label as many states as you can before checking your answers.

Name: ______________________ Date: ____________

Directions: Read the text, and study the photo. Then, answer the questions.

Electoral College

People vote for president on Election Day. But, they are not really voting for the president. They are voting for electors. The electors are the ones who vote for president. This group is called the *Electoral College*. In most states, the winning candidate gets all of the electoral votes in the state. This number of votes depends on how many people live there. In 2016, Texas had 38 electoral votes. Maine had four.

Some people think the Electoral College is fair. Without it, states with smaller populations would not have much of a voice in choosing the president. The founding fathers felt voters might not be informed. They thought electors would know more about politics.

Others think the Electoral College is not fair. A candidate might win Texas by just a few votes. Yet, he or she will receive all 38 electoral votes. The other candidate might win Maine by a lot of votes, and receive four. This means a candidate can win more votes from people, but lose the election. It has happened before!

1. What is the Electoral College?

____

____

2. How does the Electoral College show population?

____

____

3. Do you think the Electoral College is fair? Use the text to support your opinion.

____

____

____

____

Think About It

Name: ______________________ **Date:** ______________

Directions: These maps show the presidential election results from 2012 and 2016. Study the maps, and answer the questions.

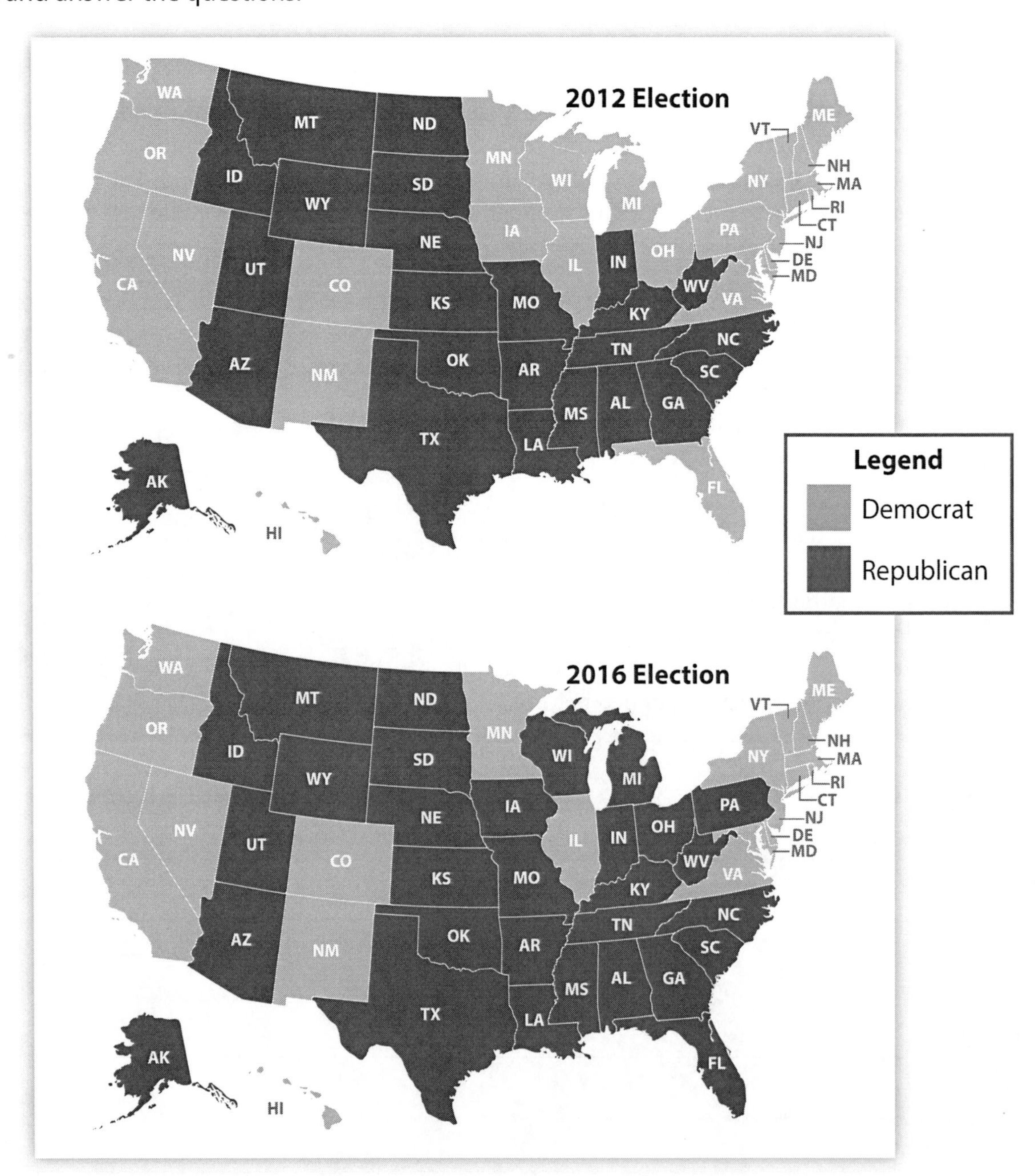

1. Name three states that changed from the 2012 to 2016 elections.

____

2. Why might some states change the way they vote?

____

____

Name: ______________________________ **Date:** ________________

Directions: Election maps show that not everyone agrees on who should be president. Write a letter to people who disagree about politics. Give advice on how they can get along.

Date: ____________________

Reading Maps

Name: ______________________________ Date: ______________

Directions: Study the map of Canadian provinces and territories. Then, answer the questions.

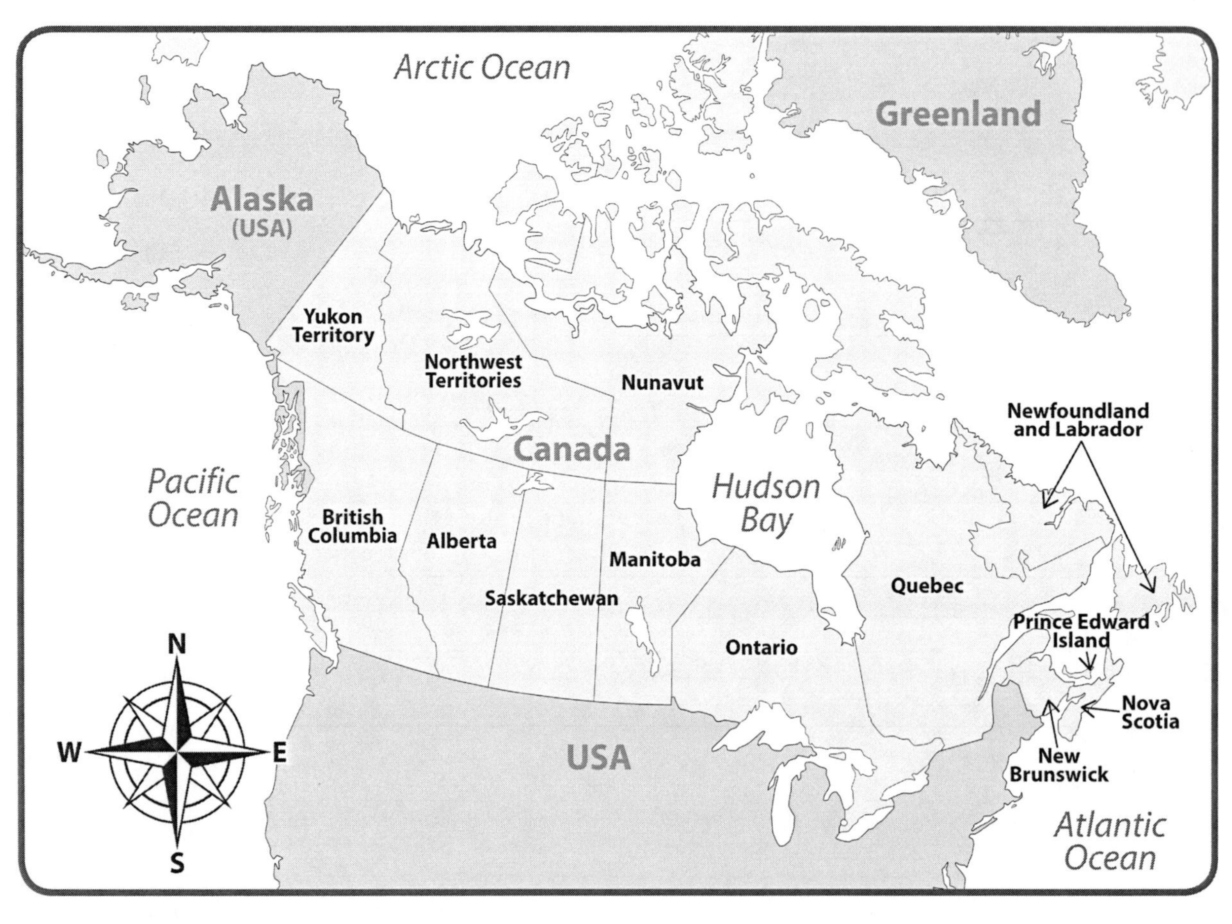

1. Which provinces border the Hudson Bay?

____

____

2. How do you think the temperatures in Nunavut compare to the temperatures in Ontario? Use evidence from the map to support your answer.

____

____

3. Which provinces do not border any ocean?

____

4. Which province extends the farthest north?

____

Name: ______________________________ **Date:** ______________

Directions: Read the text in the box. Use the information to shade the Appalachian Region on the map. Then, create a legend to show what the shaded region is.

> The Appalachian region is in the southeast part of the country. It includes all of New Brunswick, Nova Scotia, and Prince Edward Island. It includes the far eastern tip of Quebec. It also includes the part of Newfoundland and Labrador that is is close to Nova Scotia.

Legend

Read About It

Name: ______________________________ **Date:** ____________________

Directions: Read the text, and study the photo. Then, answer the questions.

The Appalachian Region

The Appalachian Mountains are among the oldest in the world. They are over 250 million years old. Most of these mountains are in the United States, but they begin in the Appalachian region of Canada. Scientists think they were formed when tectonic plates under the surface collided. That made the land buckle and fold. This process formed mountains. Water and wind have eroded the tops of the mountains over time. Many of the peaks are shorter and more rounded than those in younger mountain ranges.

The Appalachian Mountains in Canada tend to be much colder than in the south. Most of the region is forest. The soil is rich and fertile, and the trees grow very tall. The mountains are full of waterfalls, streams, and rivers. They provide water to the wildlife living there.

1. How did the mountains form?

____

____

____

2. What evidence is given that the mountains have been eroded over time?

____

____

____

3. Why are the waterfalls and streams important?

____

____

____

Name: ______________________ Date: ______________

Directions: The chart lists the heights and locations of several mountains in North America. Study the chart, and answer the questions.

Mountain Name	Mountain Range	Location	Height (in feet)
Denali	Alaska	Alaska, U.S.A.	20,310
Mt. Elbert	Rocky Mountains	Colorado, U.S.A.	14,440
Mt. Logan	St. Elias	Yukon, Canada	19,551
Mt. Mitchell	Appalachian	North Carolina, U.S.A.	6,684
Mt. Whitney	Sierra Nevada	California, U.S.A.	14,494
Pico de Orizaba	Trans-Mexican Volcanic Belt	Puebla and Veracruz, Mexico	18,406

1. Which mountain is the tallest?

____

2. Which mountain is the shortest?

____

3. Which two mountains are nearly the same height?

____

4. The Appalachian Mountains are older than the other ranges in the chart. How might that affect Mt. Mitchell's height?

____

____

____

5. What can you infer about the age of Denali in Alaska? Why?

____

____

____

Name: ______________________________ Date: ______________

Directions: Complete the Venn diagram to compare and contrast the Appalachian region of Canada to where you live.

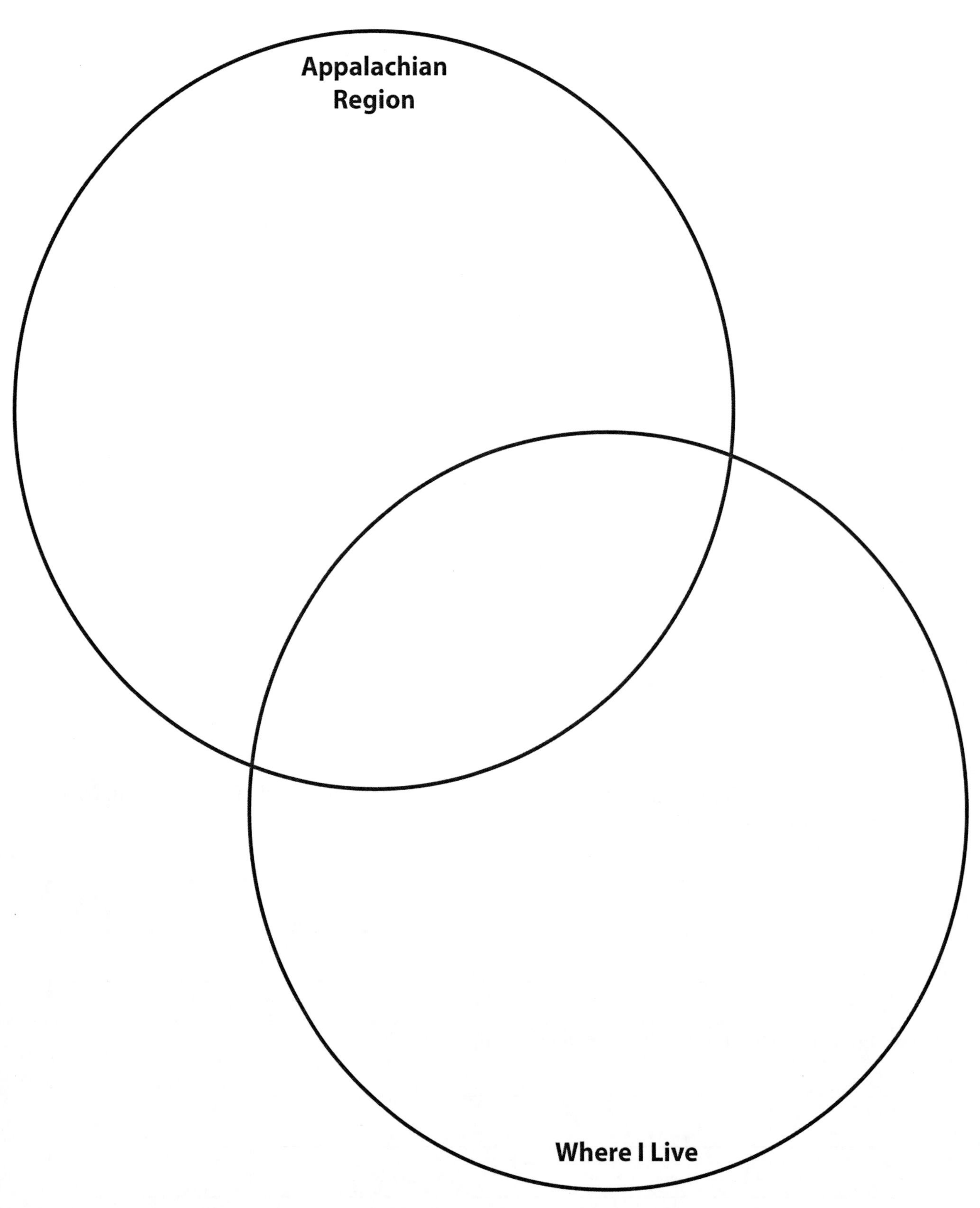

Name: ______________________ **Date:** ______________

Directions: This map shows Tornado Alley in the United States. Study the map, and answer the questions.

Tornado Alley

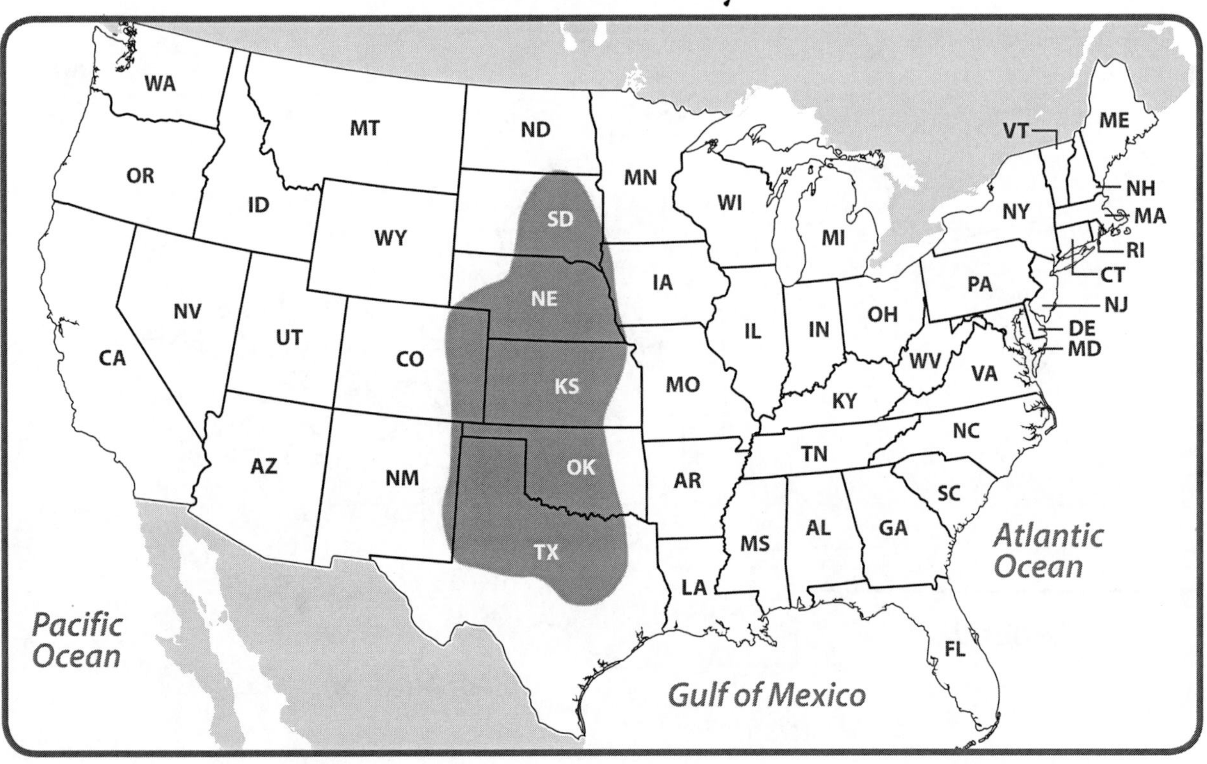

1. Which states are in Tornado Alley?

____

____

2. Why might this region be named "Tornado Alley"?

____

____

____

3. Describe the location of Tornado Alley in the United States.

____

____

____

Reading Maps

Name: ______________________ Date: ______________

Directions: Follow the steps to complete the map.

____

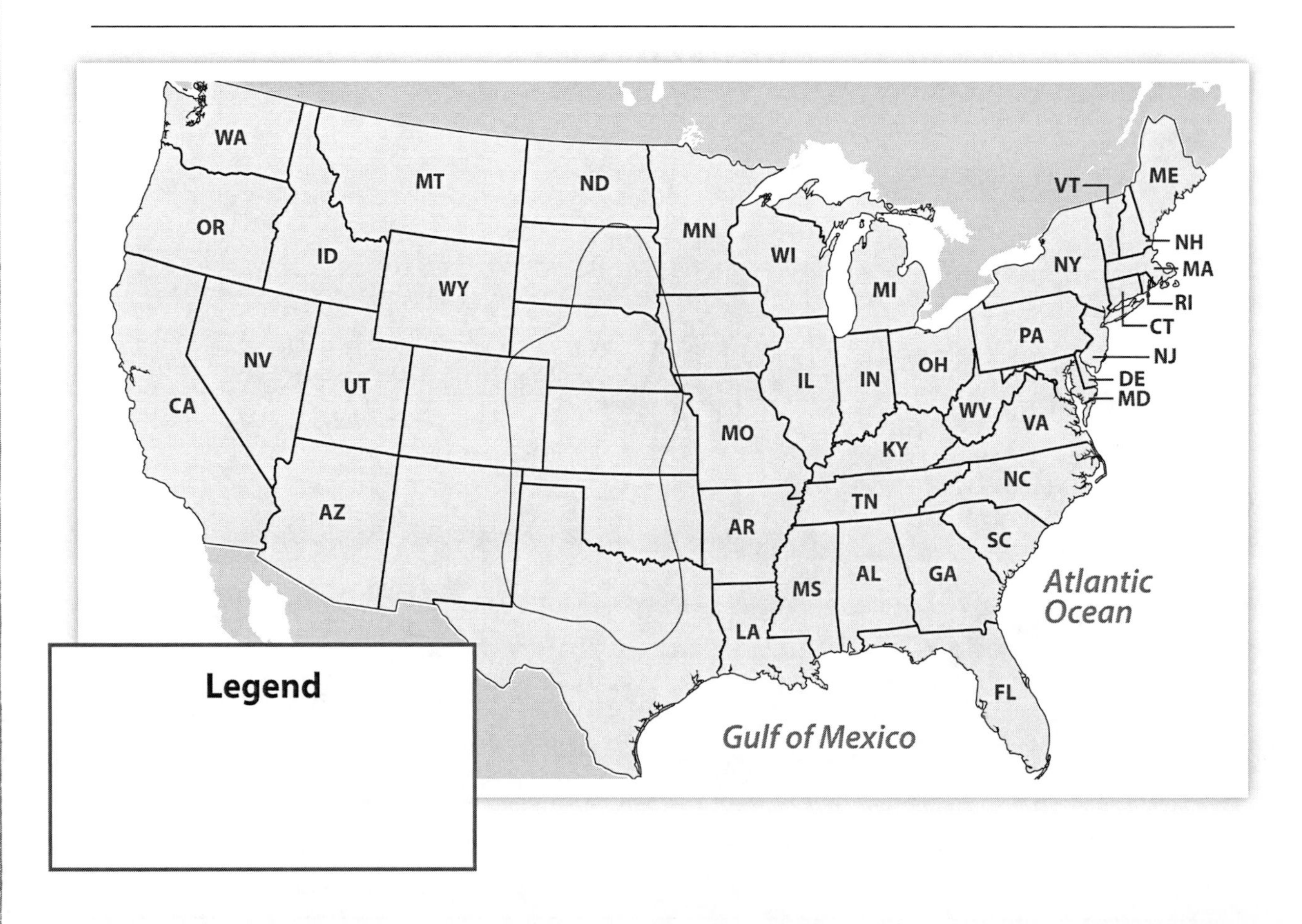

Word Bank			
Texas (TX)	Kansas (KS)	South Dakota (SD)	Colorado (CO)
Oklahoma (OK)	New Mexico (NM)	Nebraska (NE)	Iowa (IA)

1. Label the states in Tornado Alley. Use the state names in the box to help you.

2. Shade the Tornado Alley region.

3. Create a legend to show what the shaded region represents.

4. Add a compass rose to the map.

5. Add a title to the map.

Name: ______________________________ **Date:** ______________

Directions: Read the text, and study the photo. Then, answer the questions.

Tornadoes

Tornadoes take place in many countries around the world. But the United States has more than any other. Tornadoes are formed when air streams of different temperatures and moisture levels mix.

Tornado Alley is the region where most tornadoes in the United States occur. There are about 1,000 of these storms in the United States each year.

Tornadoes are rated on the Enhanced Fujita (EF) scale. The weakest is an EF-0. It has winds less than 85 miles (137 kilometers) per hour. The highest level is an EF-5. These tornadoes have winds more than 200 mph (322 kph). They cause mass destruction and can be deadly. An EF-3 is in the middle. It has winds that are 136 to 165 mph (219 to 266 kph). About 95 percent of the tornadoes in the United States are below an EF-3. A very small percentage are an EF-5. There might only be one of these tornadoes a year.

1. Describe how tornadoes are rated.

____

____

____

2. Do you think the damage to the home in the picture was caused by an EF-0, EF-3, or EF-5 tornado? Why?

____

____

____

____

3. About how many tornadoes happen in the United States each year?

____

Think About It

Name: ______________________________ **Date:** ________________

Directions: This map shows the different air streams that meet in Tornado Alley. Study the map, and answer the questions.

Tornado Formation

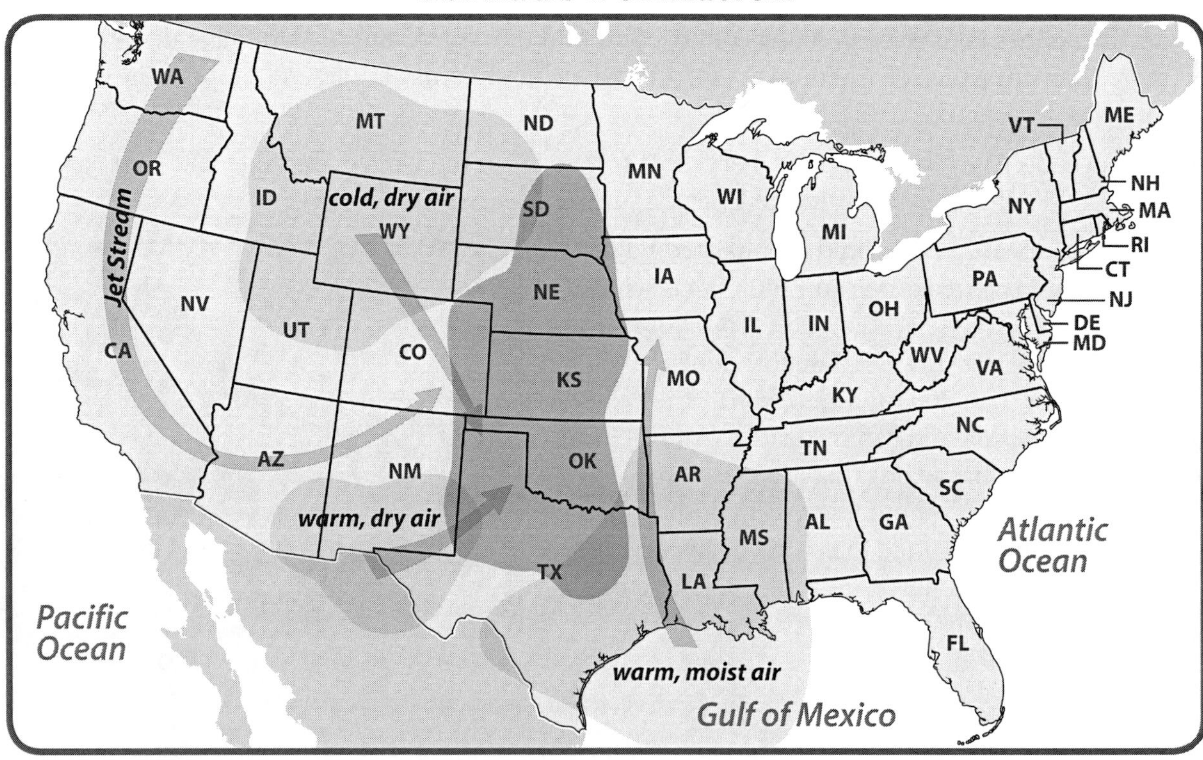

1. What types of air combine to form tornadoes?

____

____

2. Why might the air from the North be cold and dry?

____

____

3. Based on this map, why might more tornadoes occur in Tornado Alley than in other regions?

____

____

____

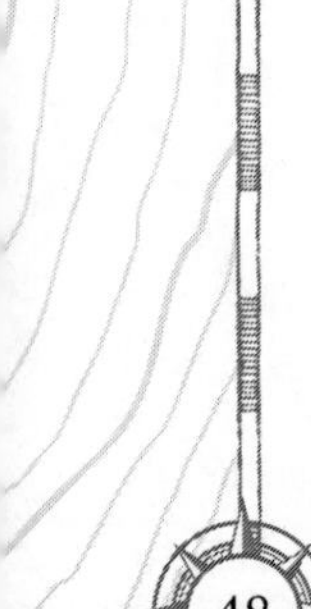

Name: ______________________________ Date: ________________

Directions: Tornadoes can destroy homes and buildings. Imagine your school has been asked to help victims of a recent tornado. Make a list of things you could do to help.

1. ____

____

____

2. ____

____

____

3. ____

____

____

4. ____

____

____

5. ____

____

____

6. ____

____

____

Reading Maps

Name: ______________________ Date: ______________

Directions: Study the map of India. Then, answer the questions.

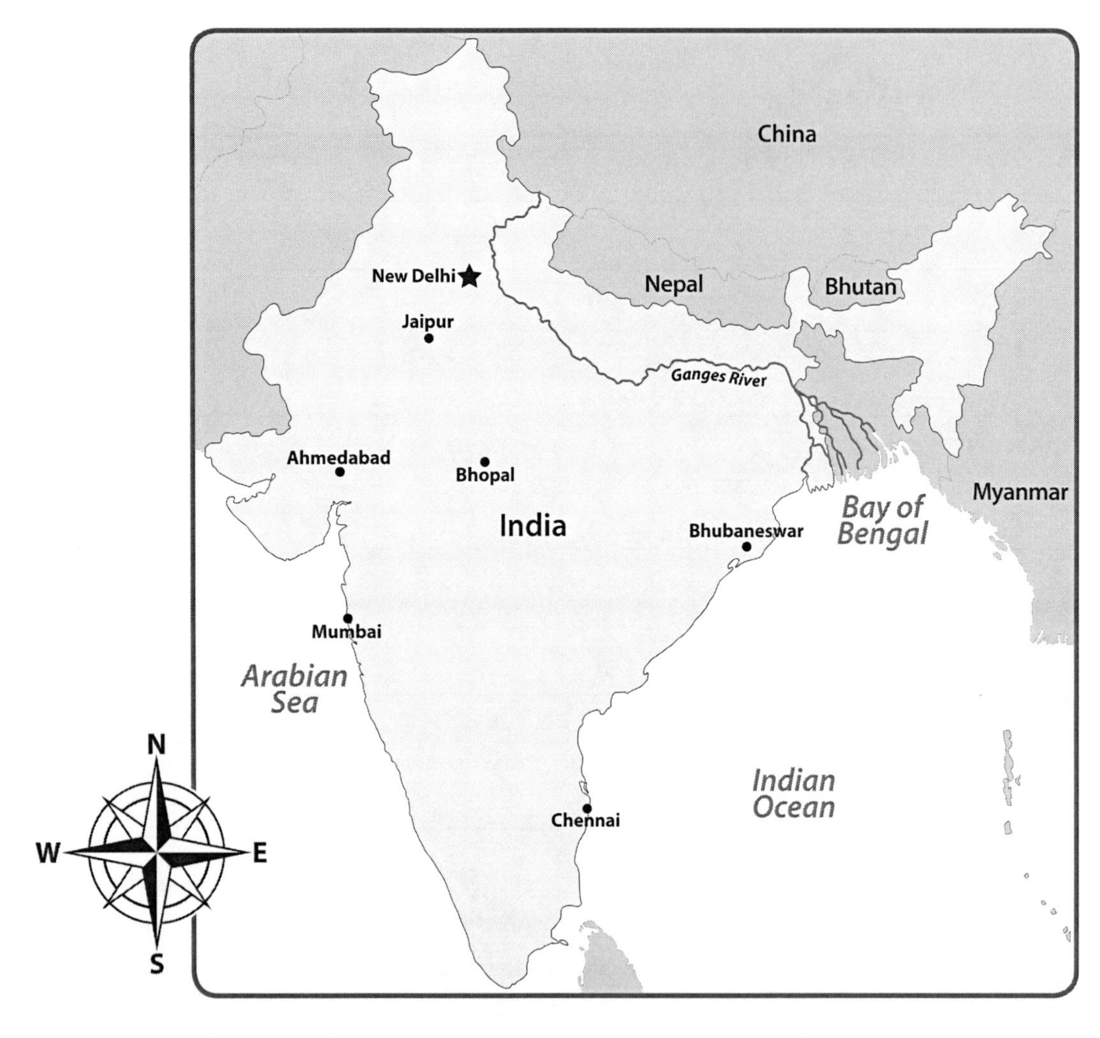

1. Find New Delhi on the map. Given its location, what might be a main source of income for this city? Why?

____

____

____

2. Which city is southwest of Bhopal?

____

3. Describe the location of the capital city.

____

____

Name: ______________________ **Date:** ______________

Directions: Follow the steps to complete the map of goods produced in India.

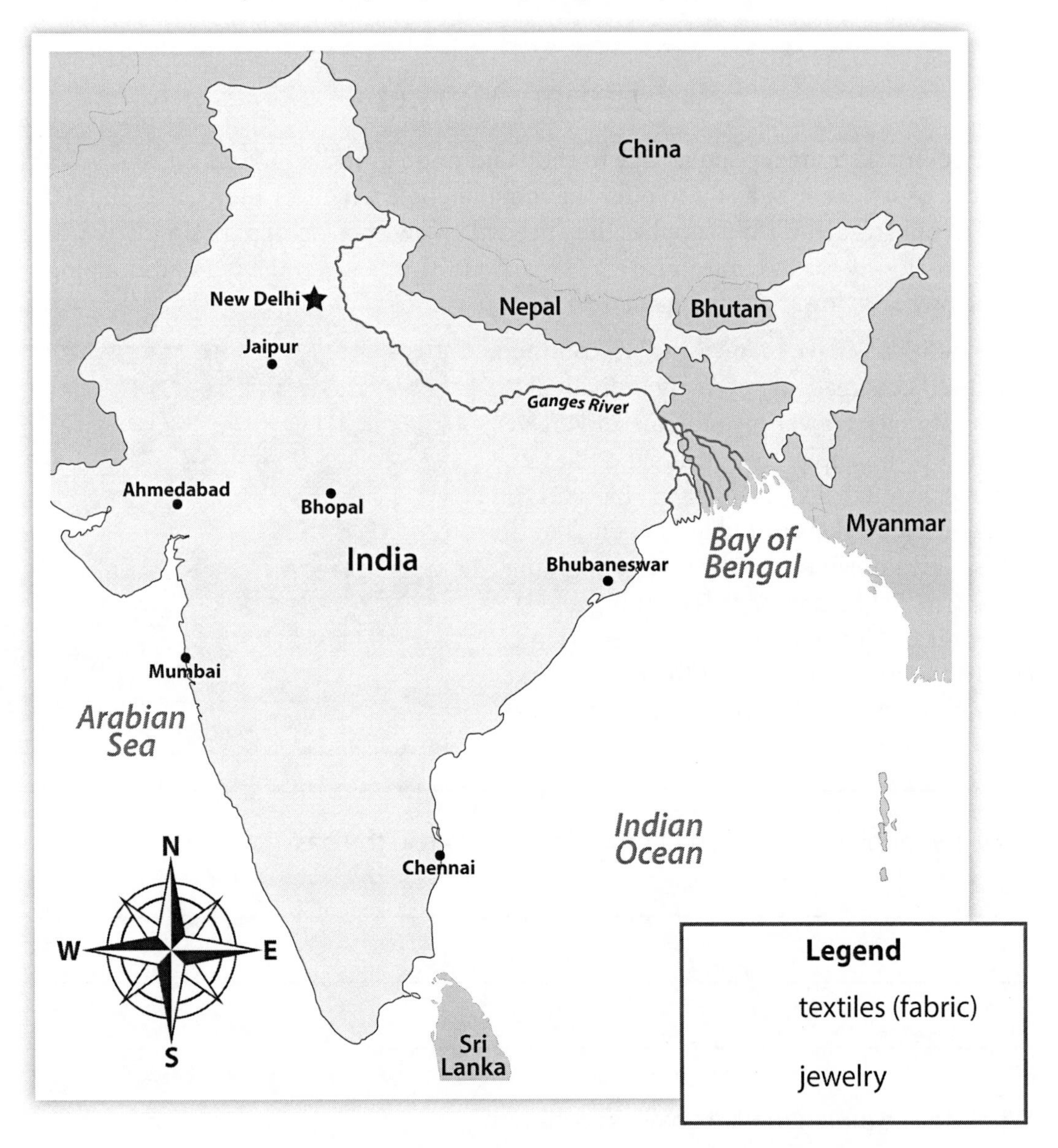

Legend
textiles (fabric)
jewelry

1. Draw a symbol next to each label in the legend.

2. Textiles are produced in Mumbai, New Delhi, and Ahmedabad. Add your textiles symbol to these locations on the map.

3. Jewelry is produced in Mumbai, Jaipur, and New Delhi. Add your jewelry symbol to these locations on the map.

Read About It

Name: 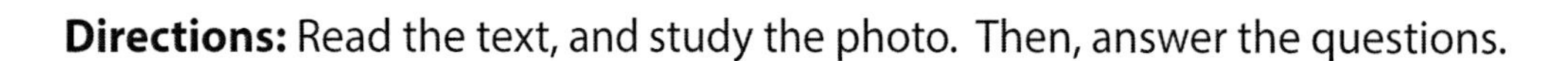 **Date:** ____________________

Directions: Read the text, and study the photo. Then, answer the questions.

Indian Markets

Street markets are a popular way to shop and trade in India. Dozens of shops cram onto the sides of streets or in alleys to offer the Indian people whatever they need. In the past, many people traded for the supplies they needed. Now, most shoppers simply pay for their items. Bartering is still common at most markets. This is when the person shopping tries to get the owner to lower the price of an item.

The markets have many things in common. Each market is made up of many individual shops. Shops sell a variety of items and are very crowded with people. Yet each market has its own special qualities. The Chawri Bazaar is one of the oldest markets in India. It opened in the 1840s. It is known for hardware and wedding cards. The Palika Bazaar sells clothing and electronics. All of its shops are underground. Wealthy shoppers might visit the Khan Market for its expensive clothes and shoes. There is no bartering at this market.

1. How is shopping at the market now different from in the past?

____

____

____

2. What do the Indian markets have in common?

____

____

____

3. Which market would you most like to visit? Why?

____

____

Name: ______________________ Date: ______________

Directions: Study these two photos of the street market in Hyderabad, India. Then, answer the questions.

1. How are these two photos the same?

____

____

____

2. How has the market changed over time?

____

____

____

3. What do you think caused these changes?

____

____

____

Think About It

Name: ______________________ Date: ______________

Directions: Fewer people are trading goods in India today. Instead, items are sold for a price and paid for with money. Think about a time you traded something and a time you bought something. Then, answer the questions.

1. Draw and write about a time you made a trade.

____

____

2. Draw and write about a time you bought an item.

____

____

3. Do you prefer to trade for an item or buy it? Why?

____

____

Name: ______________________ Date: ______________

Directions: Study the map of Israel. Then, answer the questions.

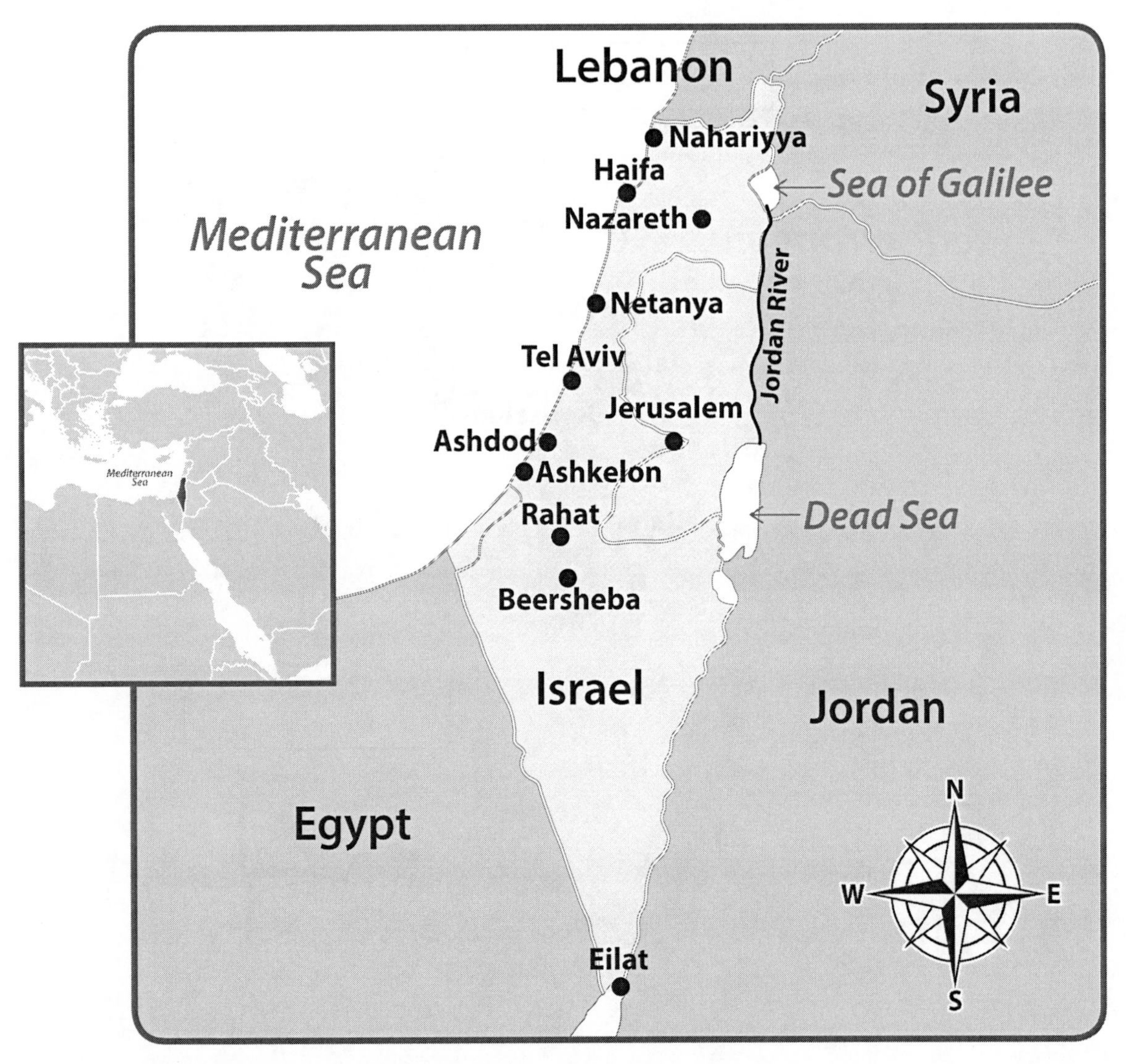

1. Name two bodies of water that make part of Israel's border.

____

2. Where are most of the cities located? Why do you think this is the case?

____

____

____

3. Name two countries that border Israel.

____

Creating Maps

Name: ______________________ Date: ______________

Directions: Follow the steps to complete the map.

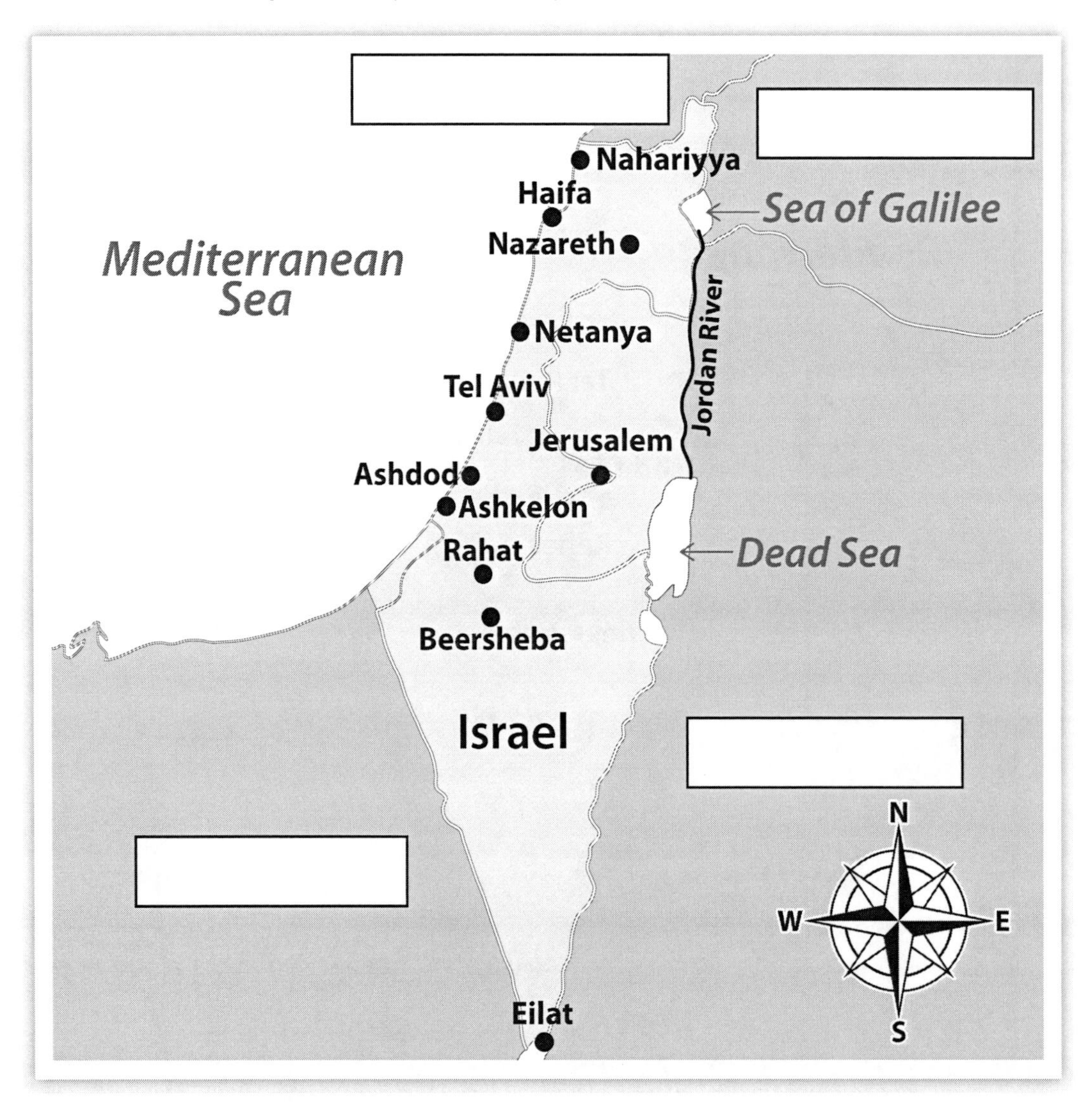

1. Most people in Israel are Jewish. But Jerusalem, Rahat, and Nazareth are home to many Muslims as well. Circle these cities on the map.

2. Color Israel green.

3. Use the clues in the box to label the countries that border Israel.

- Egypt is southwest of Israel.
- Lebanon borders Syria and the Mediterranean Sea.
- Jordan is east of the Jordan River.
- Syria is northeast of Israel.

Name: ______________________ Date: ______________

Directions: Read the text, and study the photo. Then, answer the questions.

Two Holy Sites

Israel is mostly made up of Jewish and Muslim people. The city of Jerusalem is very important to both groups. Two holy sites in the city are the Western Wall and the Dome of the Rock shrine.

Jewish people believe the Western Wall is part of a temple. The temple was destroyed, but the Western Wall remained. Thousands of Jewish people visit it each year and mourn the loss of the temple. For this reason, the wall is sometimes called the Wailing Wall.

Less than a mile from the Western Wall is the Dome of the Rock, a Muslim shrine. Muslim people believe the prophet Muhammad stood on a rock and rose to heaven. The shrine is built on that particular rock. The Dome of the Rock is not a place to worship but a place for believers to visit and remember.

1. Why do people sometimes call the Western Wall the Wailing Wall?

____

____

2. What is important about the site the Dome of the Rock was built on?

____

____

3. Why is Jerusalem an important city to both Jewish and Muslim people?

____

____

4. Which holy site would you rather visit? Why?

____

____

Think About It

Name: ______________________________ Date: ____________________

Directions: Study the graph, and answer the questions.

Religions in Israel

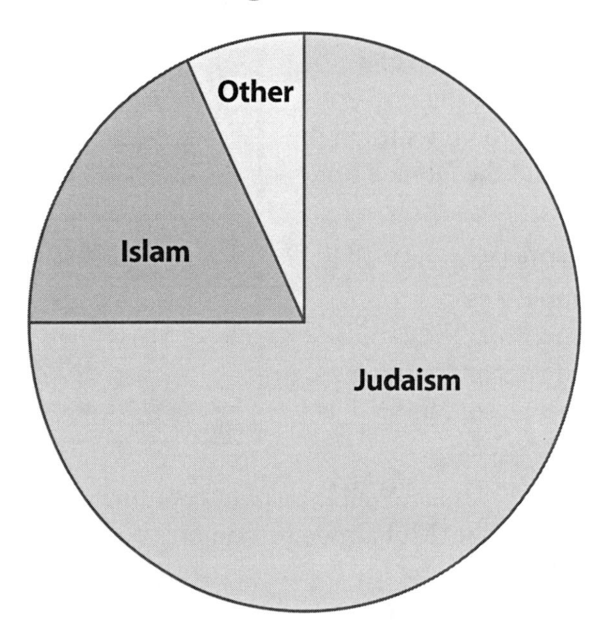

1. What religion is the largest in Israel?

____

2. What religions might be in the "other" category?

____

3. Israel is home to many religions. What conflicts may arise from this?

____

____

____

4. How might Israel benefit from being home to many religions?

____

____

____

____

Name: ______________________________ Date: ______________

Directions: In Israel, Jewish and Muslim people live together. How are people in your community different from you? What advantages and disadvantages do these differences bring? Write a paragraph to explain.

Reading Maps

Name: ______________________ Date: ______________

Directions: Study the map of Tokyo. Then, answer the questions.

Points of Interest in Tokyo

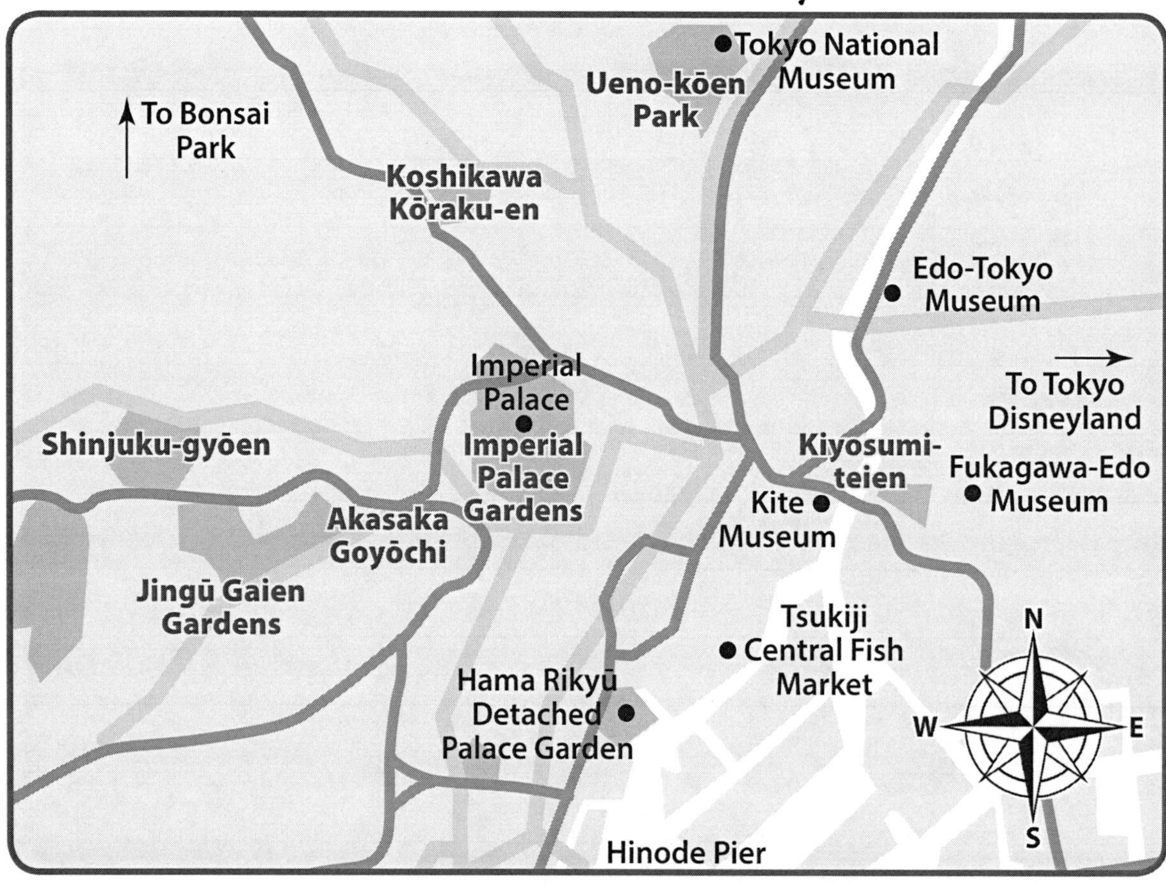

1. Describe the location of the Imperial Palace.

____

____

2. Draw a route someone might take to go from the Tokyo National Museum to the Imperial Palace.

3. Which two attractions are near this area but not shown on the map? How do you know?

____

____

4. Which attraction shown on the map would you most like to visit? Why?

____

____

Name: ______________________________ Date: ____________________

Directions: Draw a map of a fictional city. Include streets, homes, shops, and fun places to visit. Include a legend to help readers understand your map.

Name: ______________________ Date: ______________

Read About It

Directions: Read the text, and study the photo. Then, answer the questions.

Japanese Homes

Over time, homes have become more modern. Japanese homes are no exception, but they still retain parts of their traditional styles.

To let in light, Japanese homes use wooden shoji panels. They use translucent paper to let in light from outside or another room. The floors are covered in tatami mats. They are woven together from rice straw and are soft and comfortable. Since the Japanese do not wear shoes inside, the mats last for many years. Tables have very short legs and are close to the floor. People sit on thin pillows rather than tall chairs.

Japanese homes have flexible floor plans. Instead of solid walls and permanent doors, homes have panels that can slide around. They can make larger or smaller spaces as needed.

1. What makes Japanese floorplans so flexible?

____

____

2. What are tatami mats?

____

____

3. What are shoji panels?

____

____

4. Why do you think Japanese homes keep some of the traditional styles?

____

____

____

Name: ______________________ Date: ______________

Directions: The graph shows how Tokyo's population has changed over time. Study the graph, and answer the questions.

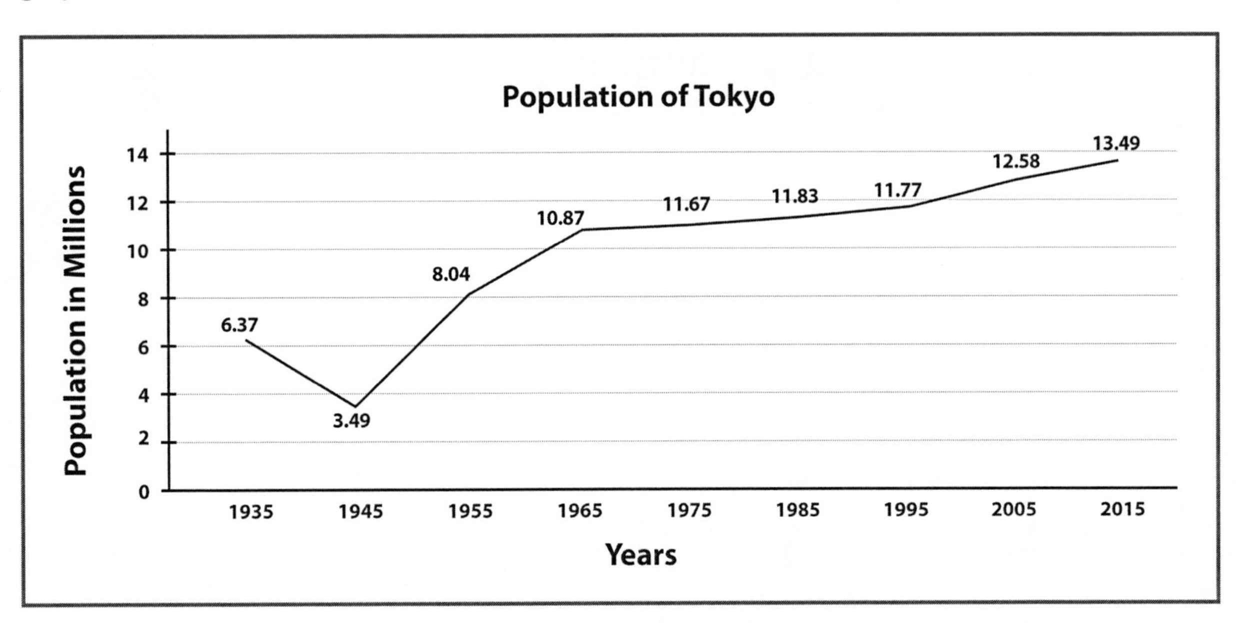

1. Which 10 years had the biggest growth in population?

____

2. Describe the overall trend of this graph.

____

____

____

3. Based on the graph, what do you predict the population will be in 2025? Why?

____

____

____

4. What effect do you think the growing population has had on housing in Tokyo?

____

____

____

Name: ______________________________ Date: ____________________

Directions: Use the chart to compare and contrast your home to a Japanese home. Then, summarize your observations.

Japanese Home	My Home

____

____

____

____

Name: ______________________________ Date: ______________

Directions: The map shows how many people live in different parts of China. Study the map, and answer the questions.

Population Density in China

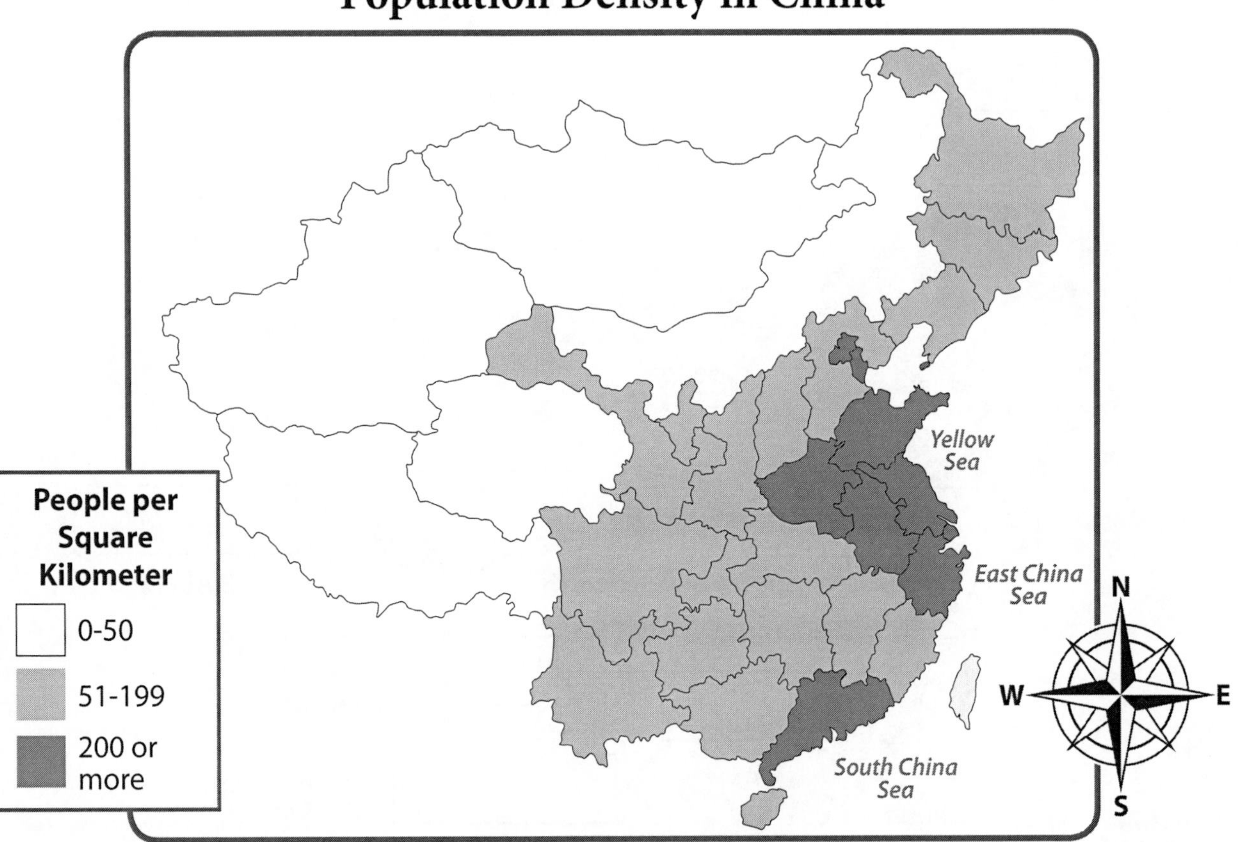

1. Describe where the highest populations are in China.

____

____

2. What bodies of water make China's eastern border?

____

____

____

3. Do you think the capital city, Beijing, is in eastern or western China? Why?

____

____

____

Creating Maps

Name: ______________________________ Date: ______________

Directions: Use the clues to label the cities on the map.

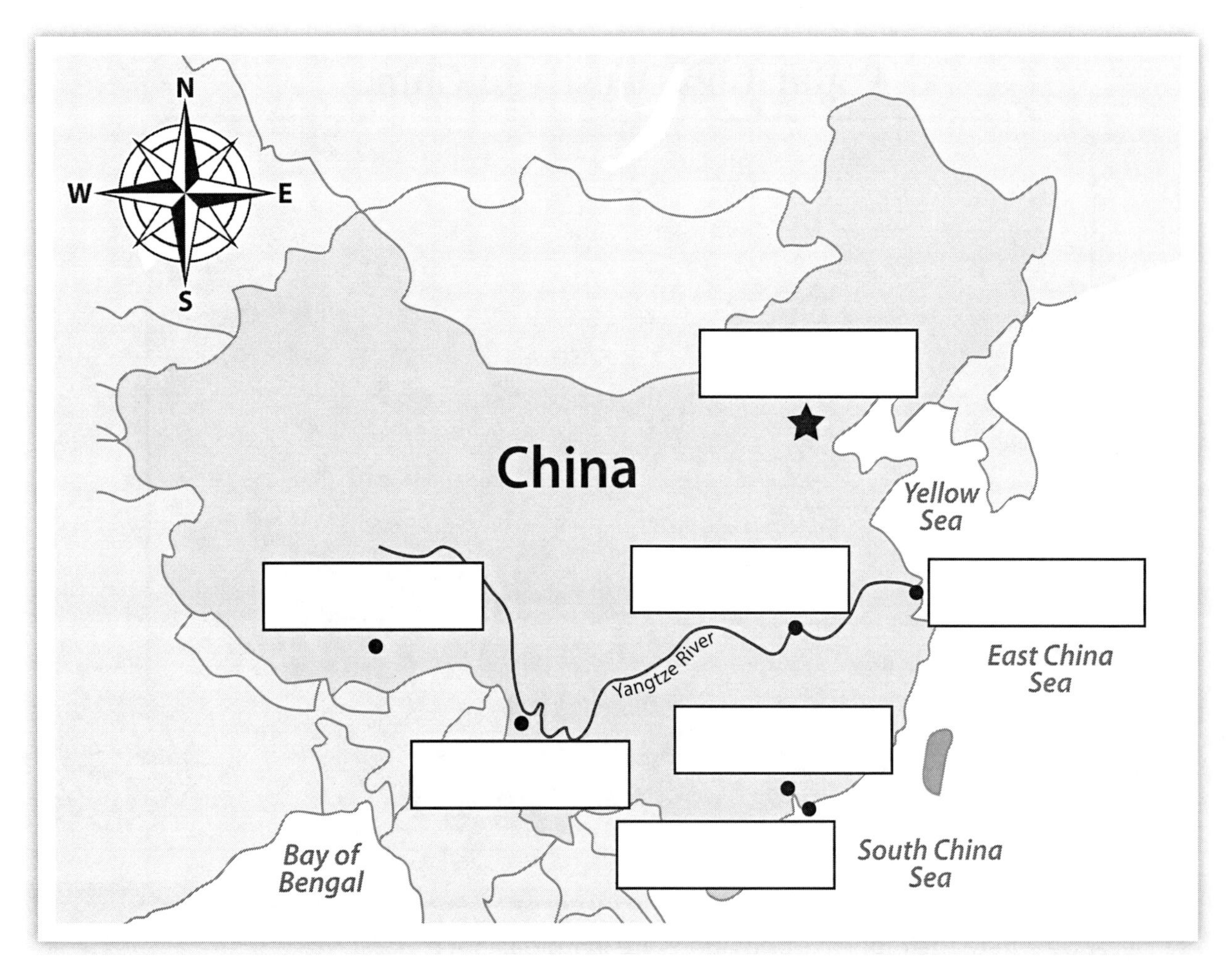

1. Beijing is the capital of China.
2. Wuhan is on the Yangtze River.
3. Hong Kong is in southern China on the South China Sea.
4. Shanghai is located where the Yangtze River meets the East China Sea.
5. Guangzhou is located just northwest of Hong Kong.
6. Lhasa is the westernmost city on this map.
7. Lijiang is southeast of Lhasa.

Name: ______________________ Date: ______________

Directions: Read the text, and study the photo. Then, answer the questions.

One-Child Policy

By the 1970s, China's population was growing very quickly. Leaders were afraid the country could not provide for so many people. They came up with a plan to slow the growth in population. Their idea was called the one-child policy. It said that Chinese couples would only be allowed to have one child. The policy became official in 1980.

Couples with only one child were rewarded for following the rules. They were given better jobs and more money. Those who did not follow the rules had to pay large fines. Some couples who had a second child tried to hide him or her. In China, having a son was important. He would carry on the family name. So, some couples would give their first-born daughters to orphanages.

The one-child policy did help slow population growth in China. But it upset many people. Leaders ended the policy in 2016.

1. Why did China begin the one-child policy?

____

____

____

2. Why would a family want to follow this policy?

____

____

____

3. What happened to some first-born daughters? Why?

____

____

____

Think About It

Name: ______________________ **Date:** ______________

Directions: Study the map, and read the text. Then, answer the questions.

Population Density in China

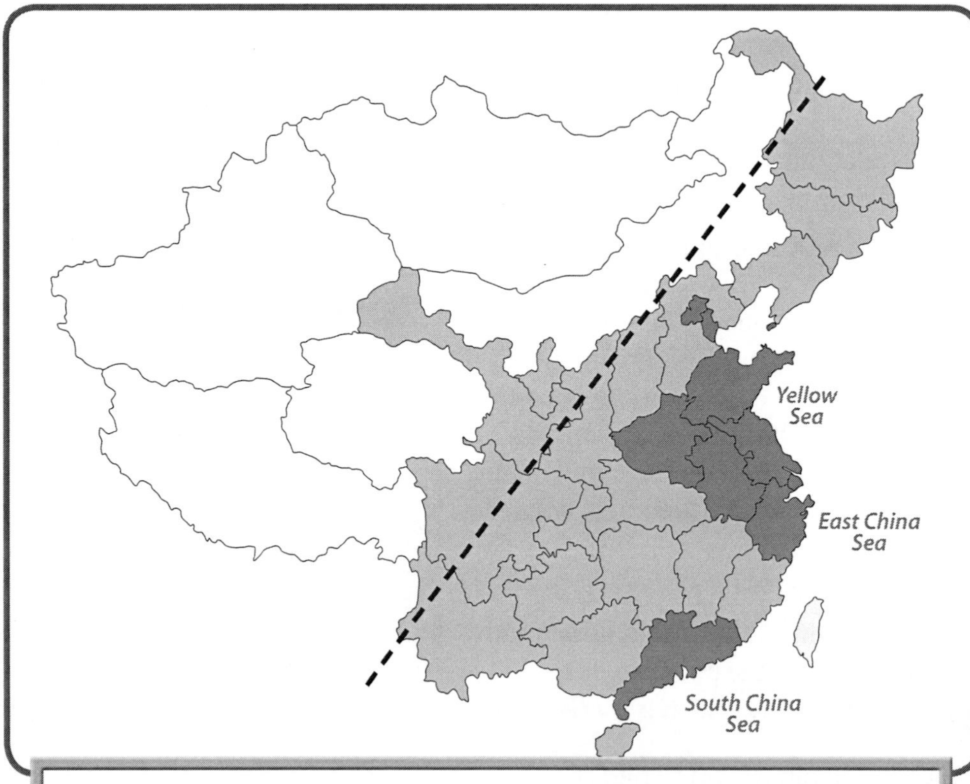

This map shows the Hu Huanyong Line. It shows the population difference between western and eastern China. About 94 percent of Chinese people live in the East. Only six percent live west of the line.

1. Why might so many more people live in eastern China?

____

____

2. How might this imbalance be a problem for both eastern and western China?

____

____

3. How might this map be related to China's decision to create the one-child policy?

____

____

Name: ____________________ **Date:** ____________

Directions: Imagine your community's population was growing very quickly. What could leaders do to help support the people? Describe a plan, and explain why you think it would work.

Geography and Me

Reading Maps

Name: ________________________ **Date:** ____________

Directions: Study the map of Southeast Asia. Then, answer the questions.

1. Which two countries are made of many islands?

____

2. Which Southeast Asian country reaches the farthest north?

____

3. Which country does not border the ocean?

____

4. Which country is at 20°N and 100°E?

____

Name: ______________________ Date: ____________

Directions: Use the information to complete the map.

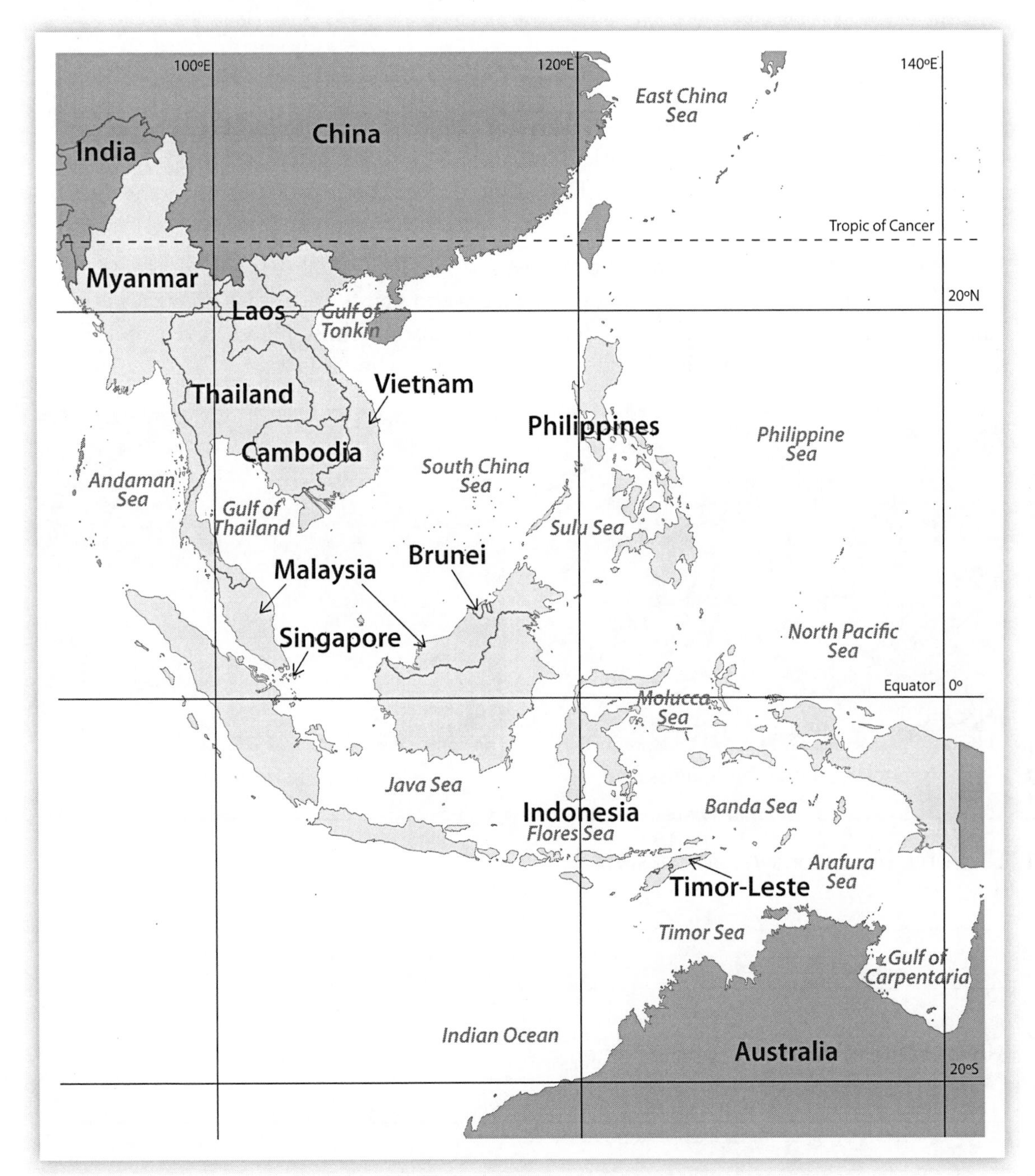

1. **Mainland countries** are connected to the rest of the continent. Shade these on the map.

2. **Island countries** are completely surrounded by water. Draw stripes on these on the map.

3. **Bridge countries** have both mainland and island areas. Circle these on the map.

Name: ______________________________ Date: ____________________

Directions: Read the text, and study the photo. Then, answer the questions.

Vegetation in Southeast Asia

The weather in much of Southeast Asia is warm and tropical. Seasons are not about warmer or cooler weather, but are marked by rainfall. In many places, there is a wet season and a dry season. The amount of rain different areas receive determines the type of vegetation they have.

Some areas do not have much of a dry season at all. It rains all year round. These places have tropical evergreen vegetation. The trees and plants are always green, and the land looks like a jungle. Other places do have a dry season. Those places have tropical deciduous forests. Trees in these forests lose their leaves when it is dry.

tropical deciduous forest in Thailand

tropical evergreen rainforest in Indonesia

1. Describe the seasons in Southeast Asia.

____

____

2. How are the forests in the photos different?

____

____

3. How might life near a tropical evergreen forest be different from near a tropical deciduous forest?

____

____

____

Think About It

Name: ______________________________ **Date:** ____________________

Directions: These photos show common crops grown in Southeast Asia. Study the photos, and answer the questions.

sugar cane field in Thailand

rice paddy in Thailand

1. Describe how each crop looks.

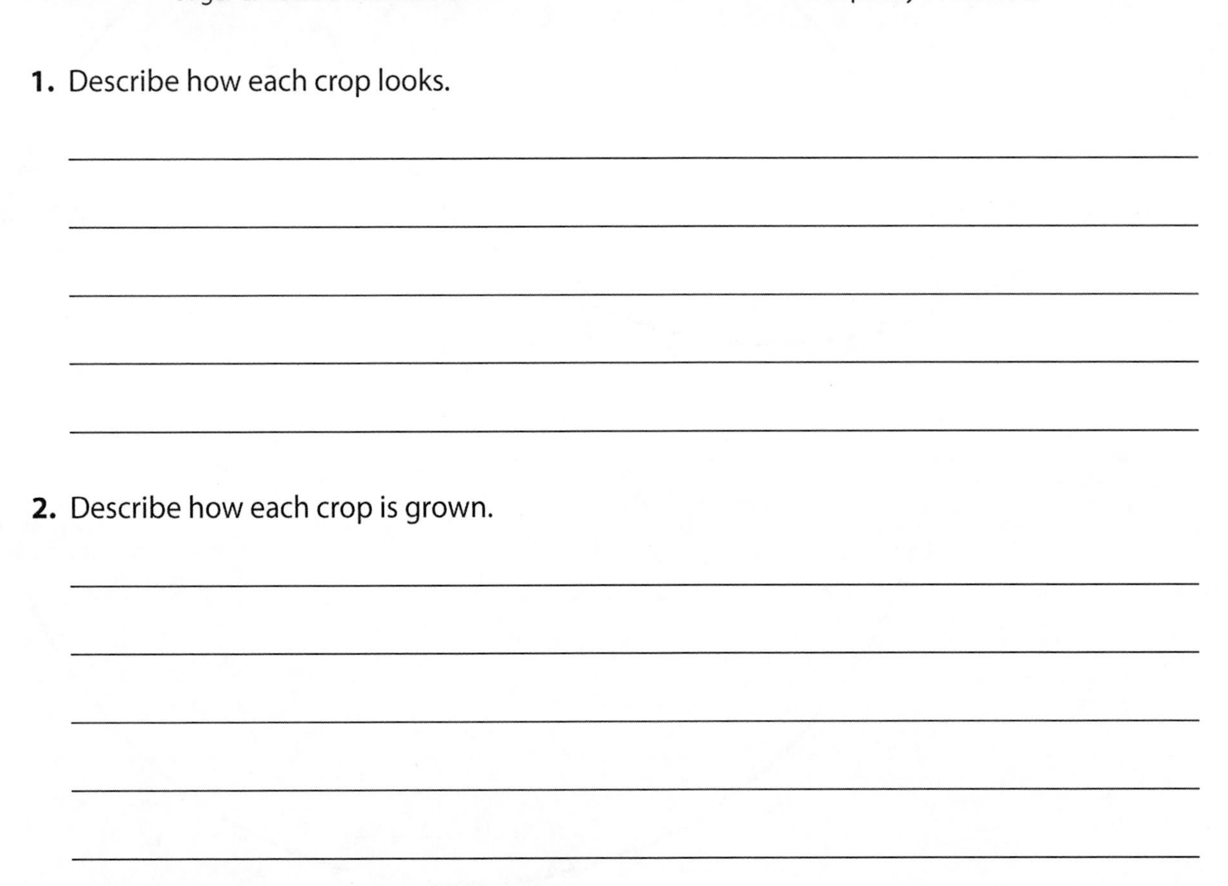

2. Describe how each crop is grown.

Name: ______________________ Date: ______________

Directions: Compare and contrast the vegetation and climate of Southeast Asia to where you live.

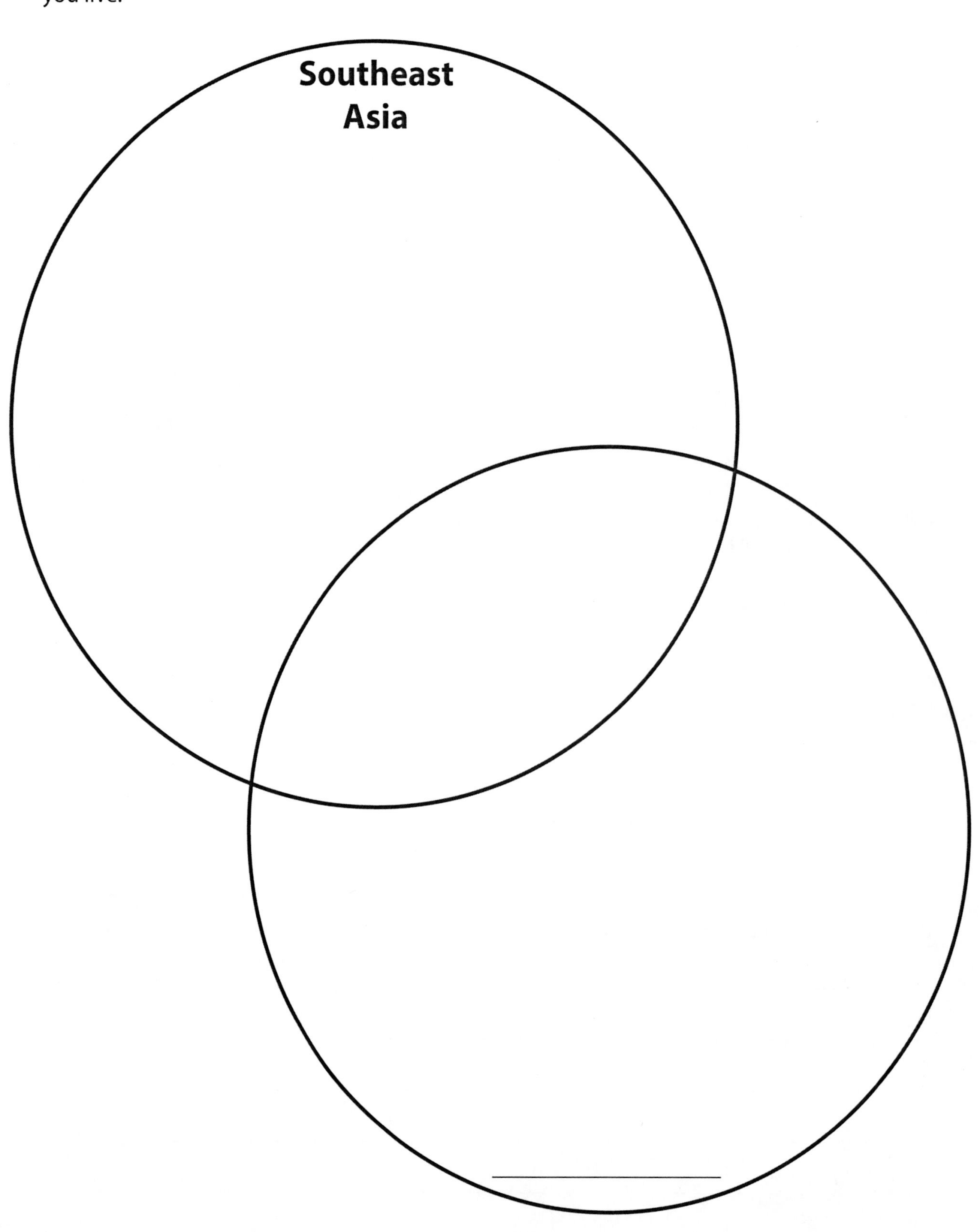

Name: ______________________________ **Date:** ______________

Directions: Study the map of Australia. Then, answer the questions.

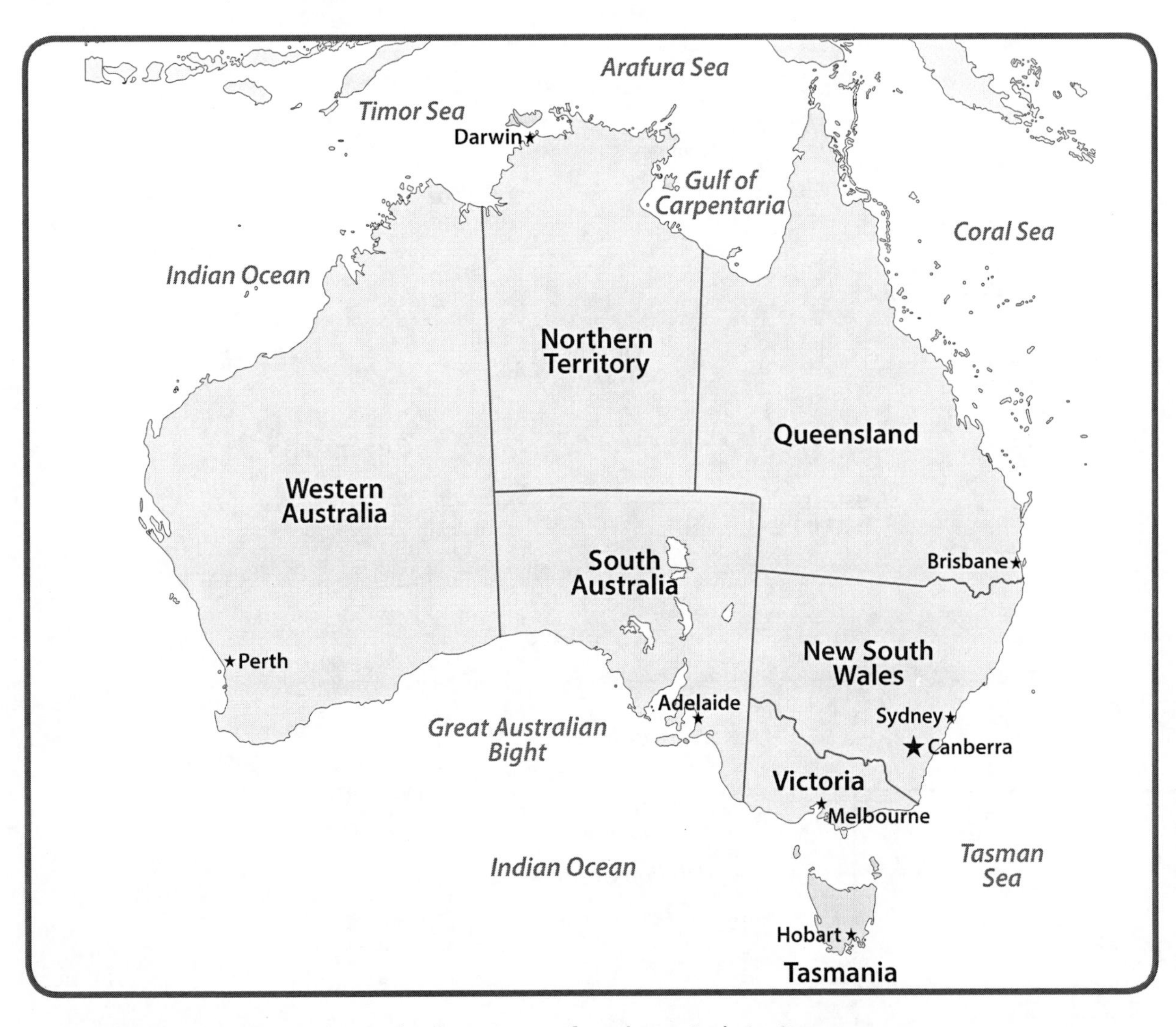

1. What do you notice about the locations of each capital city?

____

____

2. Which city is the national capital of Australia? How do you know?

____

____

____

3. How many total states and territories are there in Australia?

____

Creating Maps

Name: ______________________ Date: ______________

Directions: Follow the steps to complete the map.

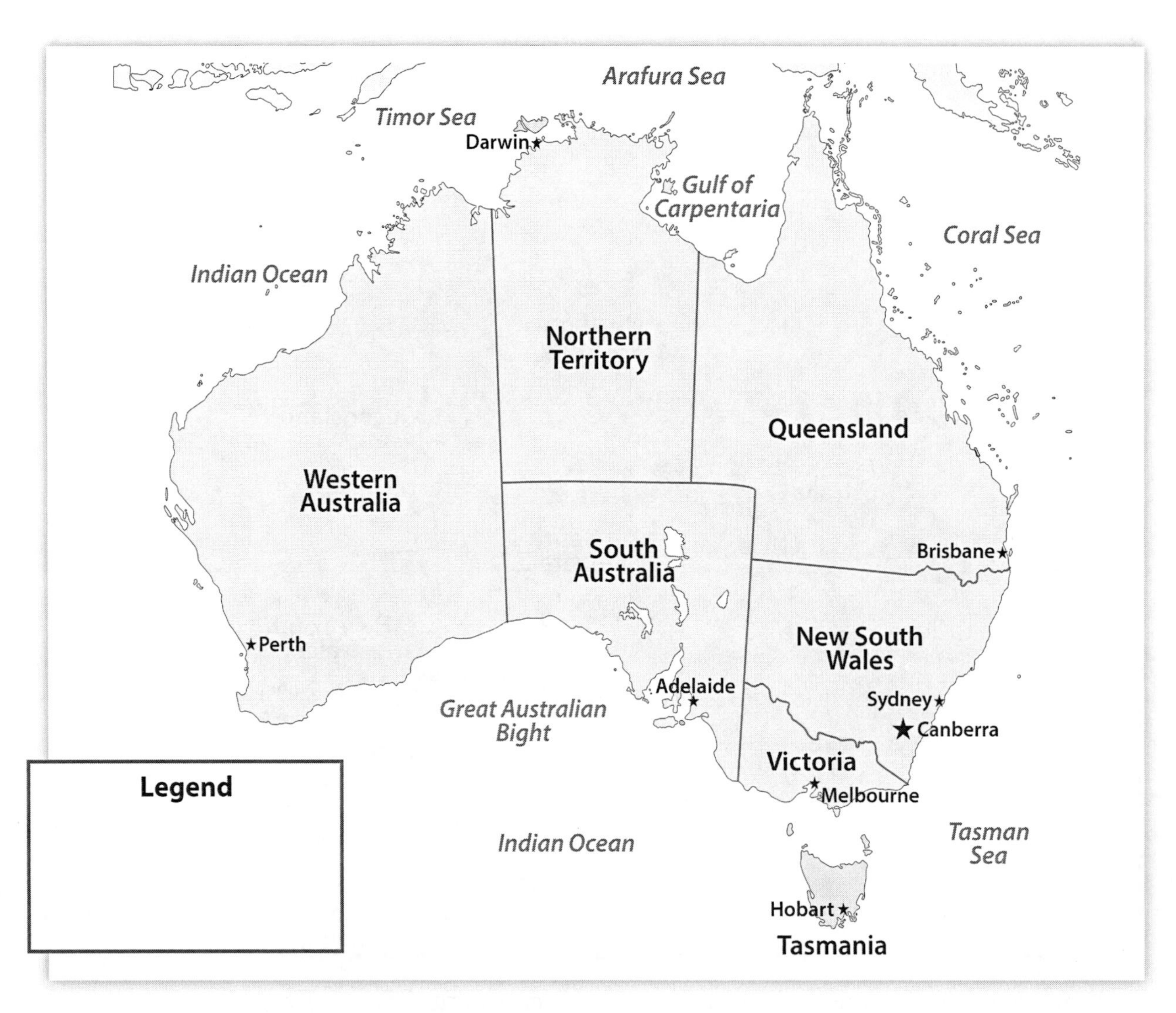

1. Draw a possible railway system for Australia. Begin at Canberra, and make the railway around the perimeter. Connect each state's capital except for Tasmania.

2. Create a legend to show what your route represents.

3. Add a compass rose to the map.

4. Shade each state and territory a different color.

5. List the bodies of water that surround Australia.

____

____

Name: ______________________________ **Date:** ______________

Directions: Read the text, and study the photo. Then, answer the questions.

Australian Railways

In the early 1800s, most people used horses to travel in Australia. But starting in the 1850s, steam engines began to change travel. The first steam engine railway was very short. It connected Melbourne to Port Melbourne. The track was only three miles long! But people knew trains were the future.

Less than 50 years later, most of the states were linked by railway. There was just one problem—the tracks were not all the same size. Private companies owned and built the railways. They made the tracks whatever size they wanted. A train could only fit on the track its company built. This meant people had to switch trains quite often.

Today, there are still three different size tracks. But rail travel is much easier than it used to be. People don't need to change trains as often as they used to. They can go from one capital to any other without switching trains.

1. How did people travel in Australia before trains?

____

____

2. What was the problem with the linked railways?

____

____

____

3. Describe the first railway in Australia.

____

____

____

Name: ____________________ Date: ____________

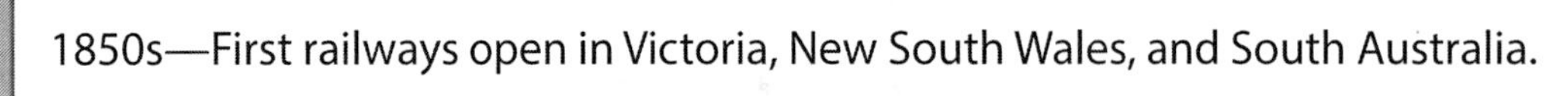

Directions: Study the timeline of railways in Australia. Then, answer the questions.

1850s—First railways open in Victoria, New South Wales, and South Australia.

1860s—Tasmania opens first railway.

1870s—West Australia and Queensland open first railway.

1880s—Northern Territory opens first railway.

1910s—Australian Capital Territory opens first railway.

1930s—Companies begin using a standard-size railway on an interstate track.

1950s—Diesel engines are introduced.

1970s—Steam engines are completely removed.

1990s—A standard-size interstate track is completed.

1. How might Australian states and territories benefit from building railways?

____

____

____

2. Why might people have wanted to build a standard-size interstate track?

____

____

____

3. Why do you think it took so long to complete the standard-size railway?

____

____

____

Name: ______________________________ Date: ______________

Directions: Think of three different types of transportation that you have used. Describe each of these forms of transportation, when they make sense to use, and why. Draw a picture of each form of transportation to support your descriptions.

Form of Transportation: ______________________

____
____
____
____
____
____

Form of Transportation: ______________________

____
____
____
____
____
____

Form of Transportation: ______________________

____
____
____
____
____
____

Name: ______________________________ **Date:** ________________

Directions: This is a map of the Northern Territory in Australia. Study the map, and answer the questions.

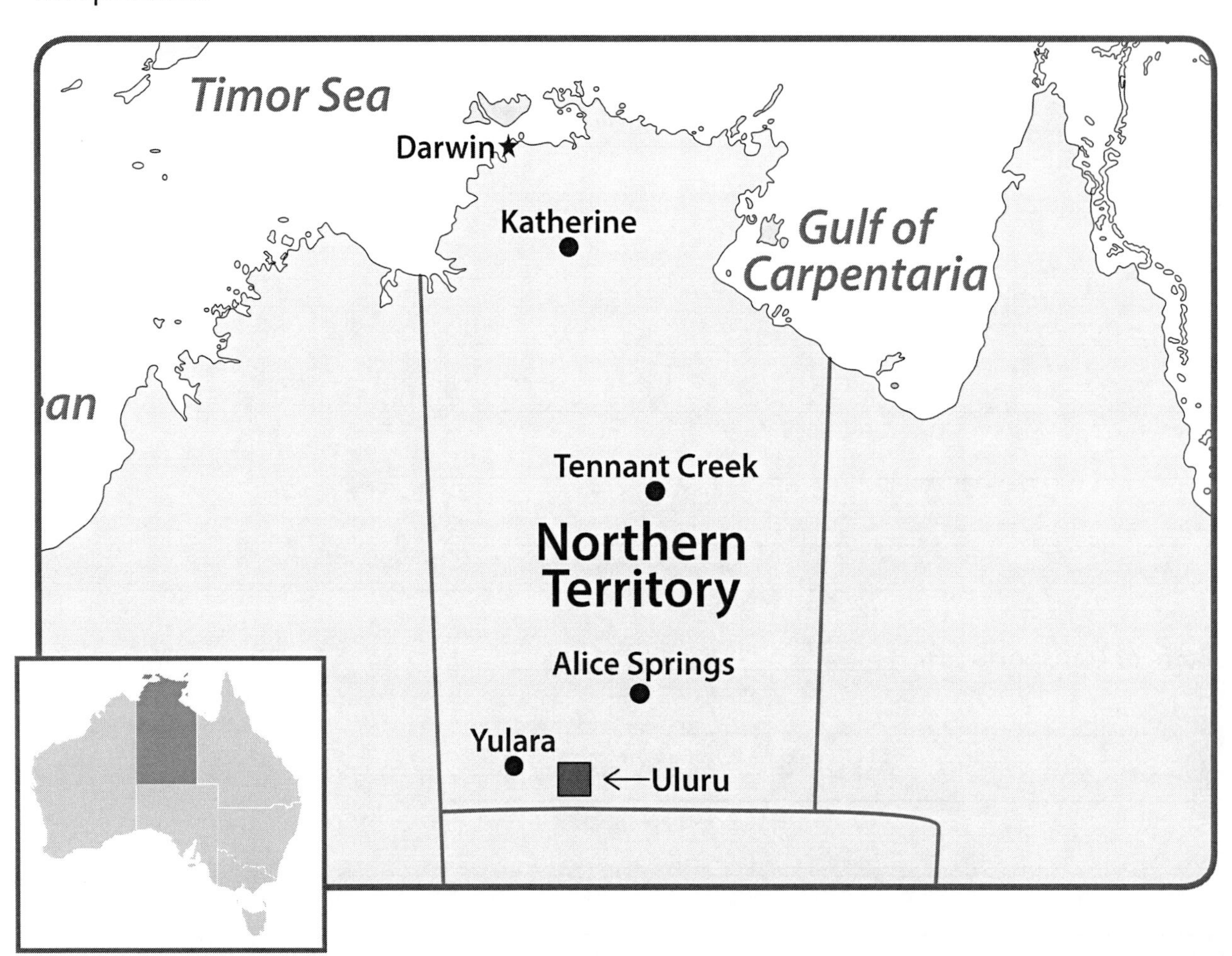

1. Which two cities are in the central part of the Northern Territory?

____

2. The capital city is on the coast of which body of water?

____

3. Uluru is not a city. What do you think it might be? What makes you think that?

____

____

____

____

Name: ______________________ **Date:** ______________

Directions: The photos below show a landform in Australia named Uluru. Use a ruler and the scales below to create scale drawings of the landform.

from the front

from above

1. Draw the front of Uluru. It is about 1,100 feet tall and 12,000 feet long.

2. Draw Uluru from above. It is 2.2 miles long and 1.5 miles wide.

Read About It

Name: ______________________ **Date:** ______________

Directions: Read the text, and study the photo. Then, answer the questions.

Uluru

One of the most majestic landforms in Australia is Uluru, also known as Ayers Rock. Uluru is its traditional name given by the Aborigines. These are the native people of Australia. The first European explorer to see the rock named it Ayers Rock.

Uluru is a monolith, which means it is a solid, single rock. It is made of sandstone. There are no plants or trees growing on it at all.

There are caves along the bottom of the rock that are sacred to the Aborigines. The rock is part of a national park, and people are allowed to climb it. But the Aborigines think it is too special to climb. They encourage people to appreciate this landform in other ways. They have a cultural center that explains the monolith's importance to their culture.

1. What is a monolith?

____

2. How did this landform get two names?

____

____

____

3. Who are the Aborigines?

____

4. Should people be allowed to climb Uluru? Why or why not?

____

____

Name: ______________________________ **Date:** ________________

Directions: Study the images. Then, answer the questions.

one of the sacred caves at the base of Uluru

Aborigine cave painting

1. What does the cave painting show?

____

____

____

2. Why might tourists want to see paintings and caves like these?

____

____

____

3. Why do you think Uluru is so special to the Aborigines?

____

____

____

____

Name: ______________________________ Date: ____________________

Directions: Think of a landform near you. Examples include mountains, canyons, valleys, rivers, deltas, and plains. Complete the chart about the landform you choose.

What is the name of the landform?	Where is it located?

Describe the landform.

Draw the landform.

How is it similar to and different from Uluru?

Name: ______________________________ **Date:** ______________

Directions: This map shows Australia's average yearly rainfall. Study the map, and answer the questions.

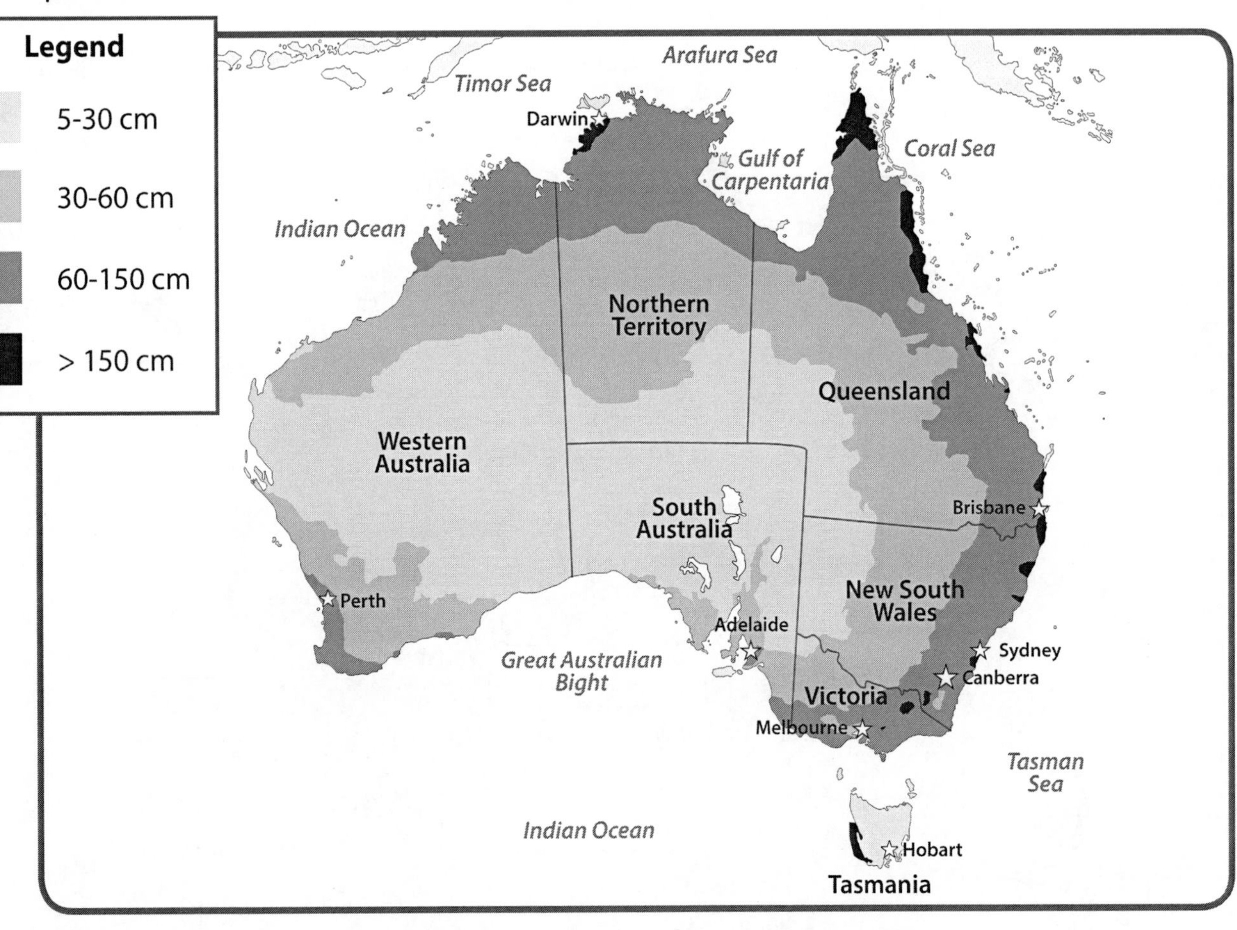

1. About how much rain does Darwin in the Northern Territory receive?

____

2. Which regions of Australia receive the least amount of rainfall?

____

3. Describe the rainfall in Tasmania.

____

____

4. How might the amount of rainfall help people decide where to create a city?

____

____

Creating Maps

Name: ____________________ **Date:** ____________

Directions: Use the clues to label the cities.

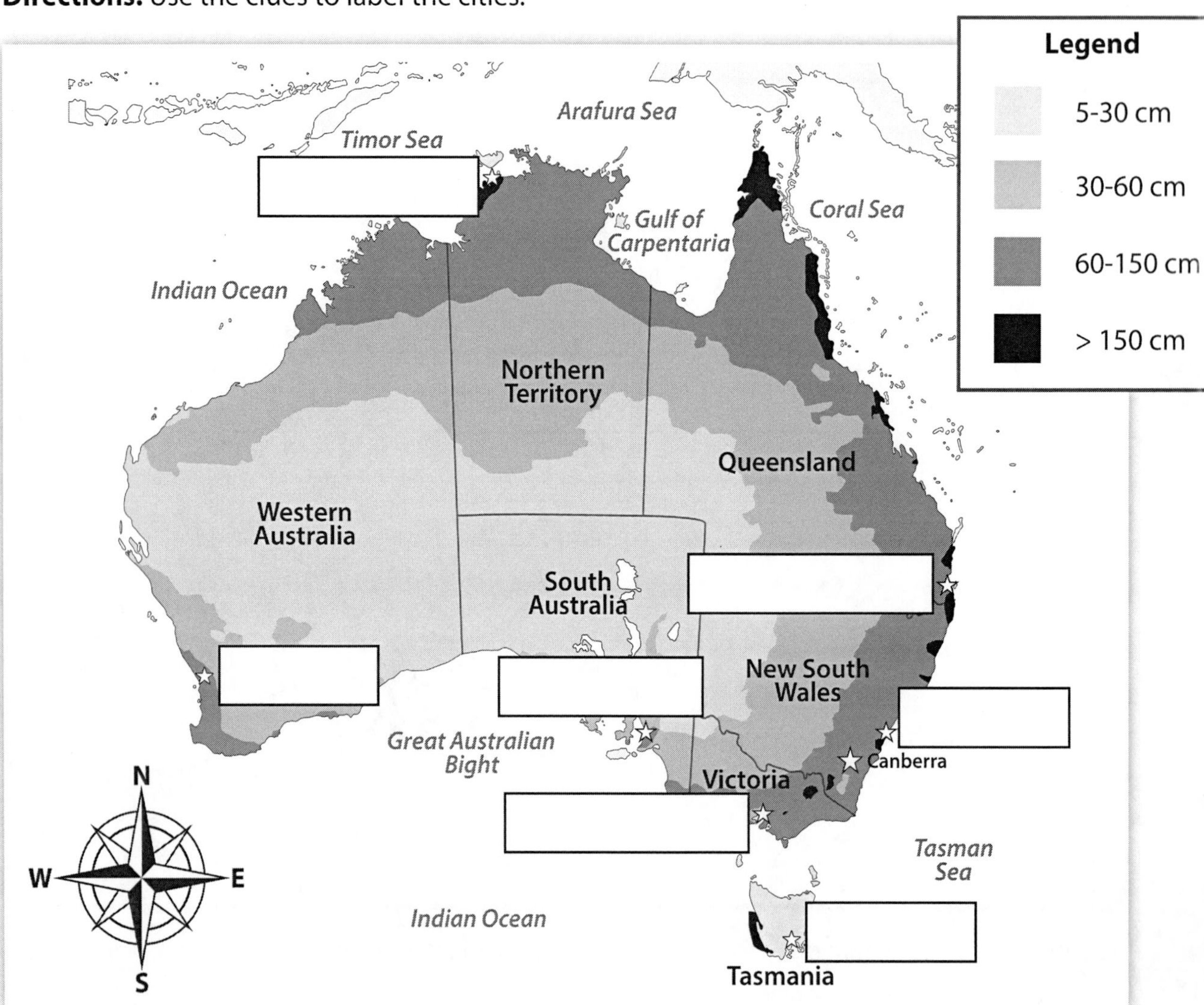

1. Hobart is on the island of Tasmania.
2. Melbourne is just north of Tasmania.
3. Perth is the farthest west.
4. Brisbane is located the farthest east.
5. Darwin receives more than 150 cm of rain each year on average.
6. Adelaide is between Melbourne and Perth.
7. Sydney is southwest of Brisbane.

Name: ______________________ Date: ______________

Directions: Read the text, and study the photo. Then, answer the questions.

Earth's Seasons

Many people know the closer a place is to the equator, the warmer it will be. But what about seasons? Many places have warm weather one month and cold weather six months later.

Earth has seasons because it is tilted on its axis. When the Southern Hemisphere is tilted toward the sun, it gets more sunlight. So, the temperatures are higher. At the same time, the Northern Hemisphere is tilted away from the sun. It gets less sunlight and has lower temperatures. When the Northern Hemisphere is tilted toward the sun, the opposite is true.

The United States and Australia are in opposite hemispheres. This means they have opposite seasons. When Australia is having summer, the United States is having winter. Christmas is in December. In the United States, this is winter. There may be snow on the ground. But that is summertime in Australia! It is often hot on Christmas.

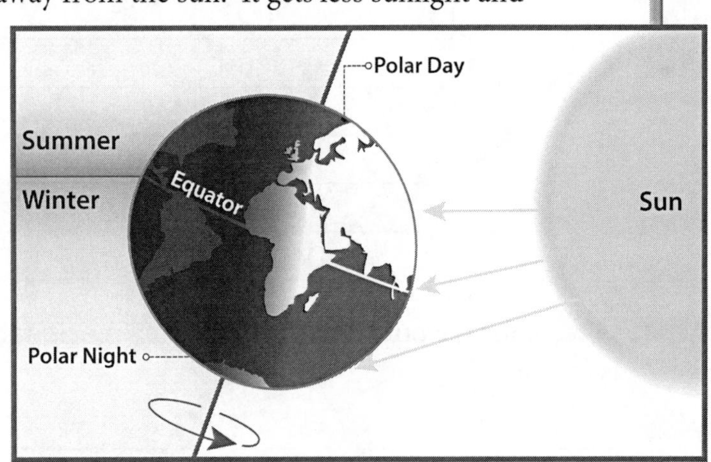

1. Why are there seasons on Earth?

____

____

2. How might Australia's location affect how people celebrate Christmas?

____

____

3. Use the picture and text to explain why the United States and Australia have opposite seasons.

____

____

____

Think About It

Name: ______________________ Date: ______________

Directions: Study the chart, and answer the questions.

Monthly Average Temperatures in Canberra, Australia	
January	81.9 °F
February	81.1 °F
March	76.1 °F
April	68.0 °F
May	60.6 °F
June	54.1 °F
July	52.7 °F
August	55.8 °F
September	61.2 °F
October	66.9 °F
November	72.7 °F
December	79.3 °F

1. In what form would most of Canberra's precipitation fall? Why? Remember, 32 degrees is freezing.

____

____

2. Which three months have the coldest temperatures?

____

3. Circle the winter months in Australia. Put boxes around the summer months.

4. How do Australia's seasons compare to the seasons where you live?

____

____

____

5. In which month would you like to visit Canberra? Why did you choose that month?

____

____

Name: ______________________________ **Date:** ______________

Directions: Imagine celebrating a holiday, birthday, or other special event during the opposite season from when you celebrate it now. How might that affect the big day? Describe the day, and draw a picture of you celebrating it in the chart.

Reading Maps

Name: ______________________________ **Date:** ______________

Directions: Study the map of New South Wales. Then, answer the questions.

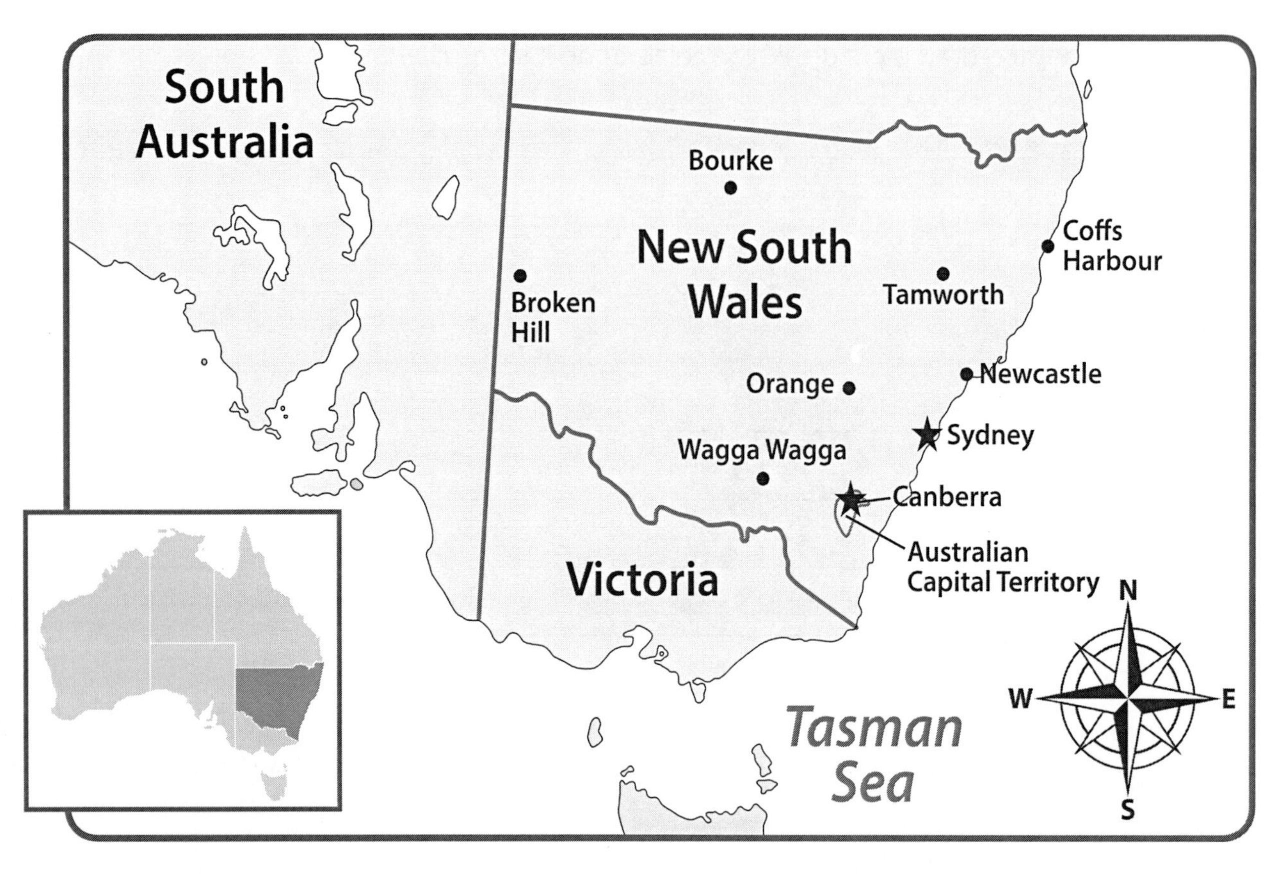

1. Name three cities on the east coast.

____

____

____

2. Which city is northeast of Orange?

____

3. Which city is the farthest north?

____

4. Where are most cities located in this Australian state?

____

Name: ______________________________ Date: ________________

Directions: Imagine you are one of the first European settlers in Australia. Create symbols for the items in the legend. Add one item of your own. Then, use your symbols to make a map of your new community.

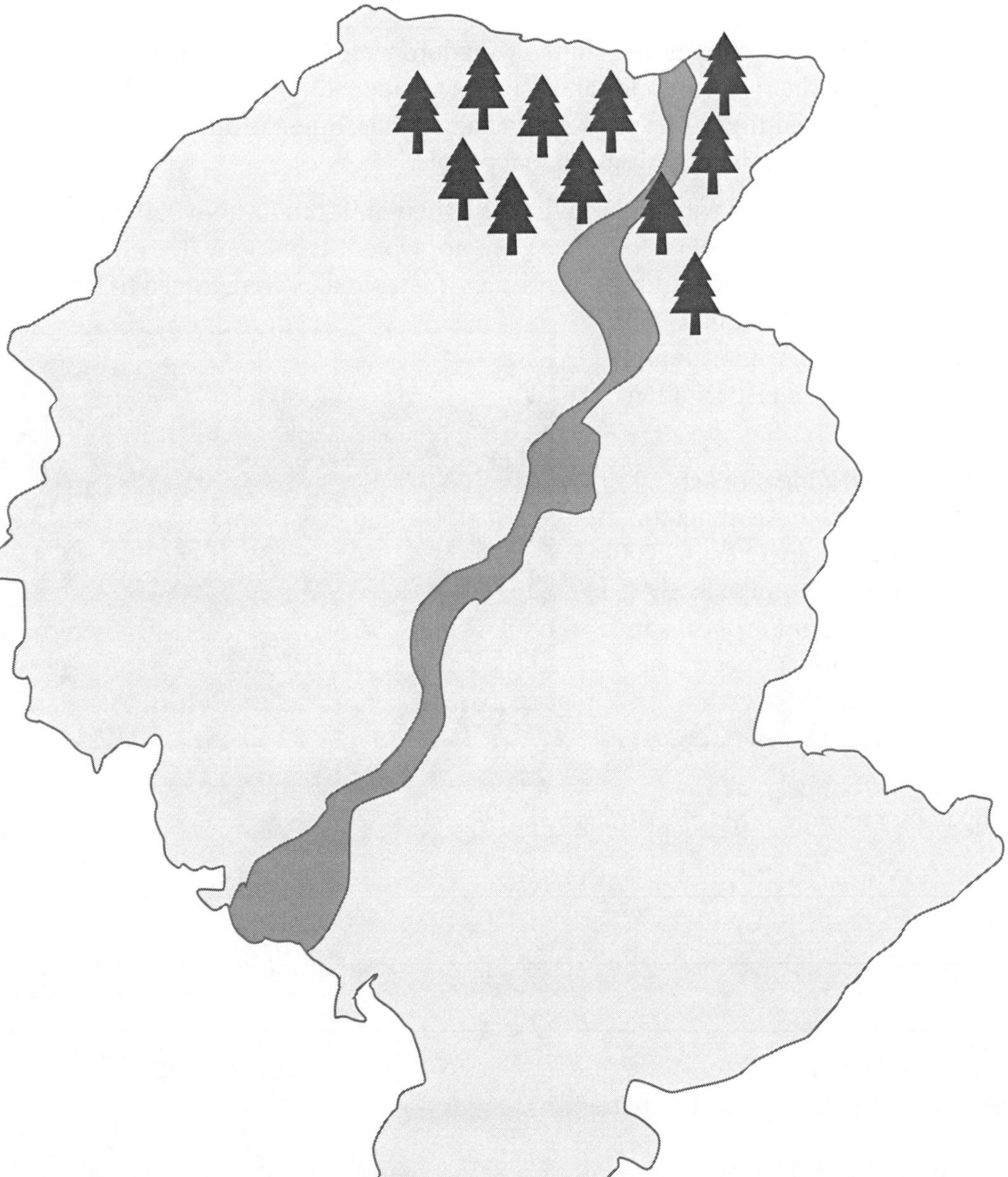

Legend

house food storage road

store bridge

Name: ______________________ Date: ______________

Read About It

Directions: Read the text, and study the photo. Then, answer the questions.

Prison Colony

In the 1700s, large cities in the United Kingdom were crowded. There were many poor people. Some of them stole money and food, but leaders did not have enough jails. So in 1788, they began sending convicts to Australia. At that time, European people did not live there. It was home to the native Aboriginal people.

The first fleet arrived with 751 convicts and their children. About 250 marines and their families also came. Most of the convicts were not violent criminals. They had stolen to survive. The convicts were not kept in prisons. Instead, they were forced to farm and build roads, bridges, and buildings. The convicts had to work many hours every day. Colonial leaders often beat them.

The United Kingdom kept sending convicts to Australia for 80 years. There were over 162,000 men and women sent in all. By the end, the number of Europeans in Australia had grown to over one million people.

1. Why were convicts sent to Australia?

____

____

____

2. What did the convicts do in Australia?

____

____

____

3. What challenges do you think the convicts faced when they arrived in Australia?

____

____

Name: ______________________ Date: ____________

Directions: Study the graph. Then, answer the questions.

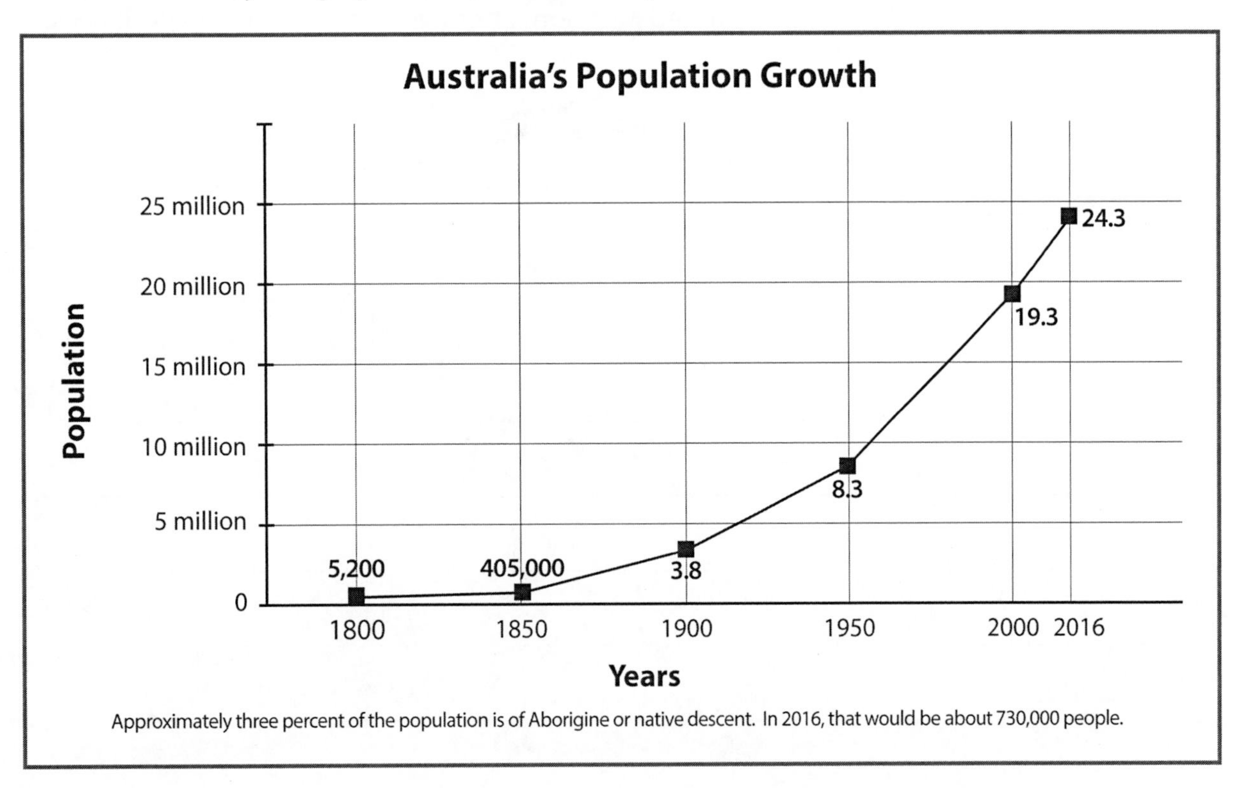

Approximately three percent of the population is of Aborigine or native descent. In 2016, that would be about 730,000 people.

1. Aborigines were not included in the official population until 1971. How would this information affect the graph?

____

____

____

____

2. Between which two years marked on the graph was the biggest growth in population?

____

3. Why do you think there are no population records before 1800?

____

____

____

____

Name: ______________________________ Date: ____________________

Directions: Imagine you are sent to a prison colony in Australia because your father is a Marine. Write a letter to a friend in the United Kingdom about your feelings and experiences.

Date: ____________________

____

Name: ______________________________ **Date:** ____________

Directions: Study the map of Australia. Then, answer the questions.

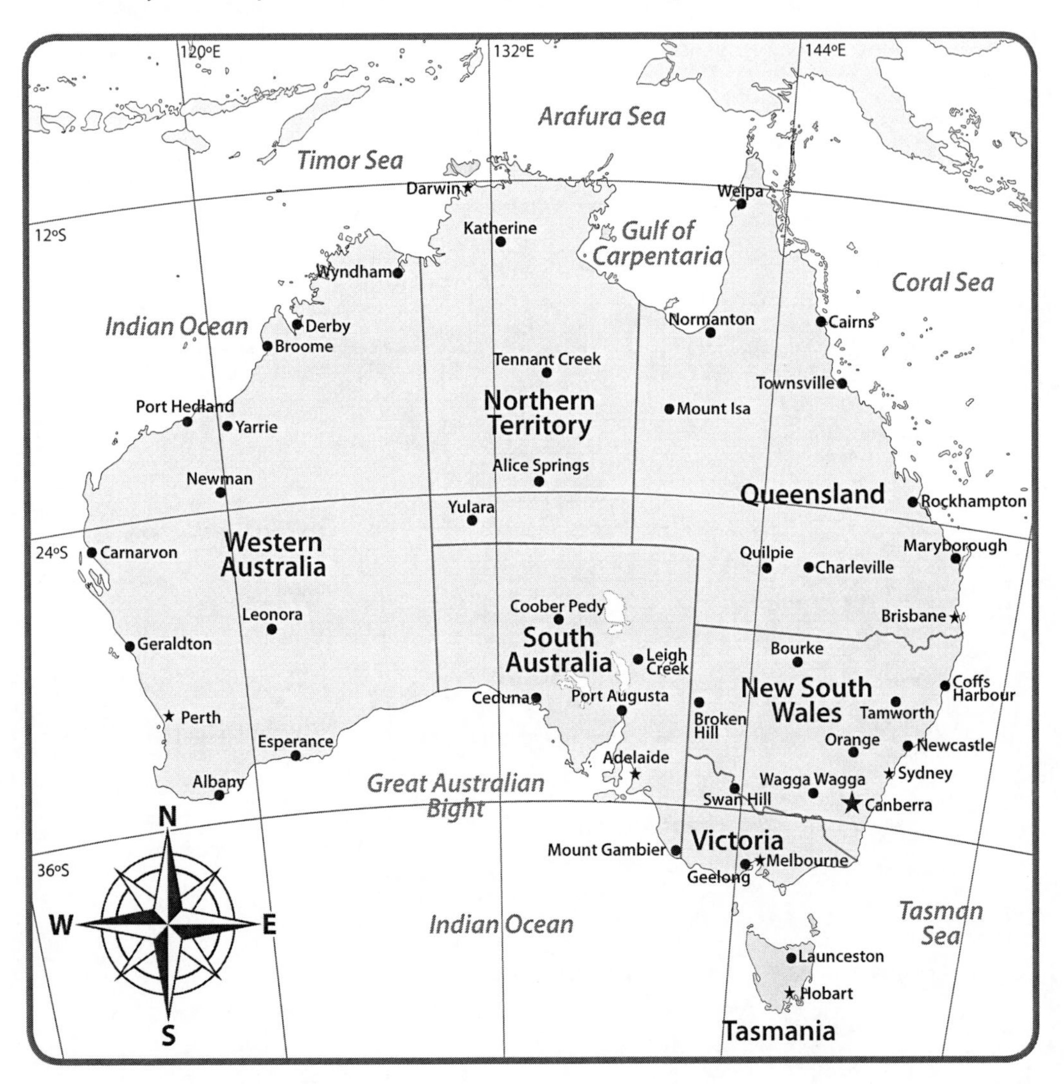

1. What city is near 30°S, 120˚E?

____

2. What is the approximate latitude and longitude of Newman in Western Australia?

____

3. Name three cities south of the 36°S line.

____

4. What is the approximate latitude and longitude of Townsville in Queenland?

____

Name: ______________________________ **Date:** ____________________

Directions: Finish the map using the latitude and longitude coordinates listed.

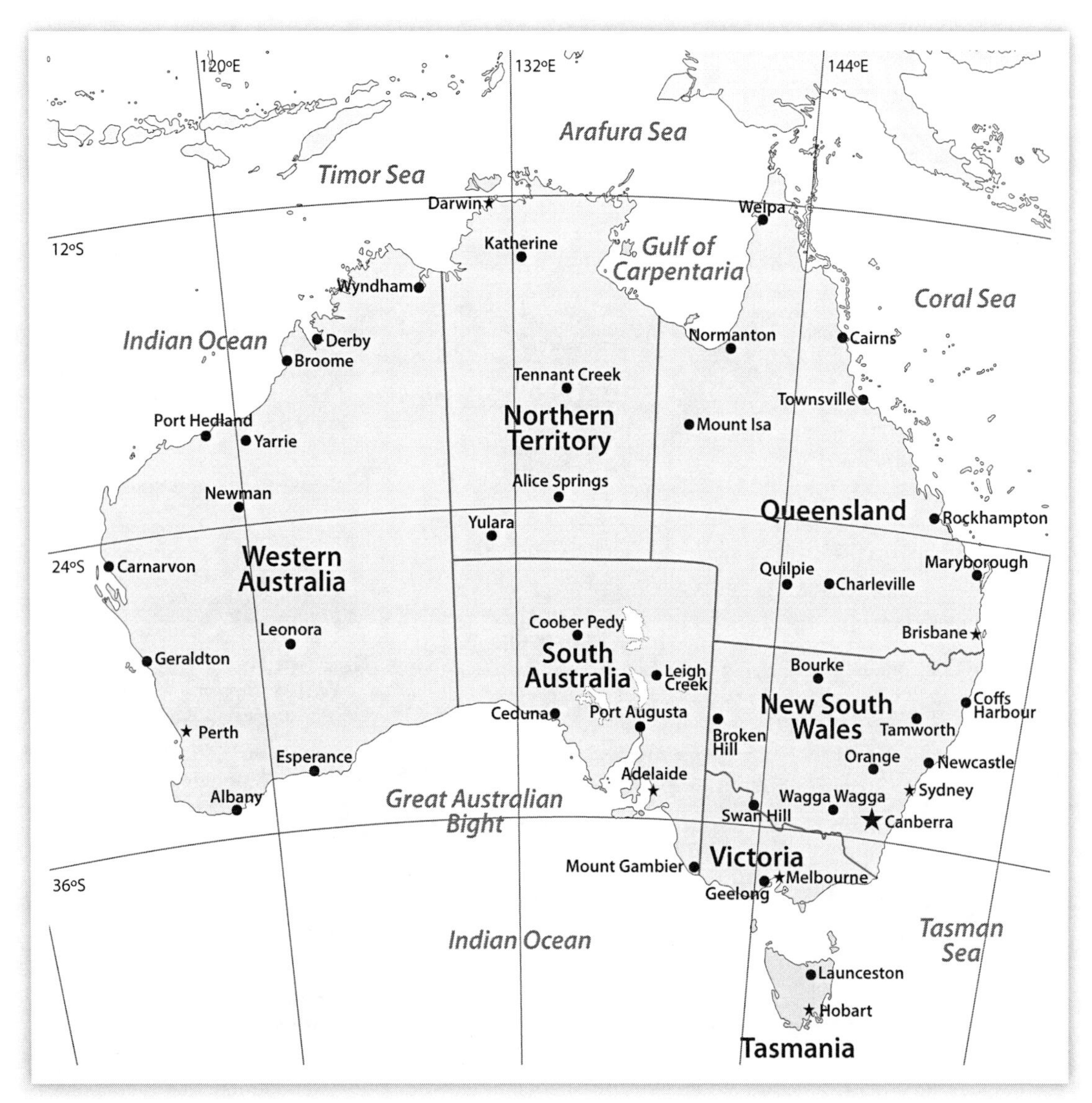

1. Write *A* at 10° S, 130° E.
2. Write *B* at 30° S, 120° E.
3. Write C at 20° S, 140° E.
4. Write *D* at 35°S, 145°E.
5. Write *E* at 15°S, 125°E.
6. Write *F* at 25°S, 135°E.

Name: ______________________________ Date: ______________

Directions: Read the text, and study the photo. Then, answer the questions.

Absolute vs. Relative Location

Imagine trying to find Sydney, Australia, on a map. There are two ways to find it. One is by its relative location. This means the city is described by referring to another location. For example, Sydney is south of Newcastle. Or it is on the east coast of Australia.

Another way to find Sydney is by its absolute location. This means its exact place on a map. Someone might use latitude and longitude to give the absolute location of a place. Latitude lines are horizontal. They measure how far north or south a place is from the equator. The equator is zero degrees. Longitude lines are vertical. They measure how far east or west a place is from the prime meridian. The prime meridian in Greenwich, England, is zero degrees. Both types of locations can help people find places.

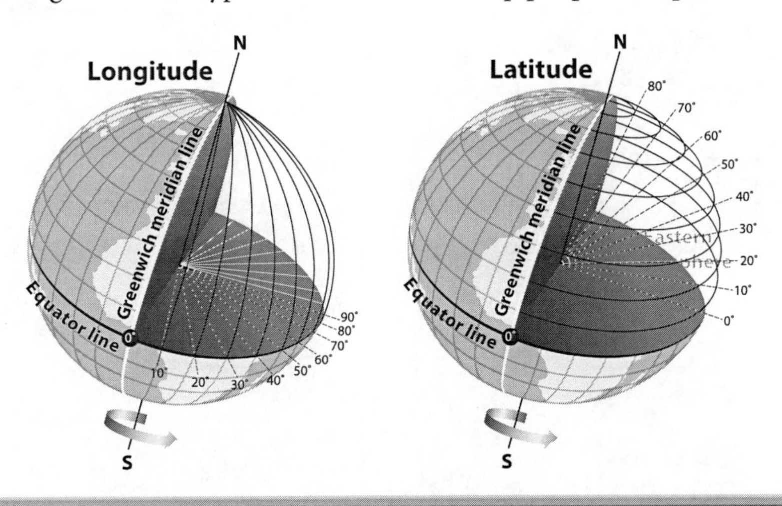

1. How are absolute and relative locations similar and different?

____

____

____

____

____

2. What marks the zero degree lines for latitude and longitude?

____

____

____

Read About It

Think About It

Name: ______________________ **Date:** ______________

Directions: This is a map of Queensland, Australia. Study the map, and answer the questions.

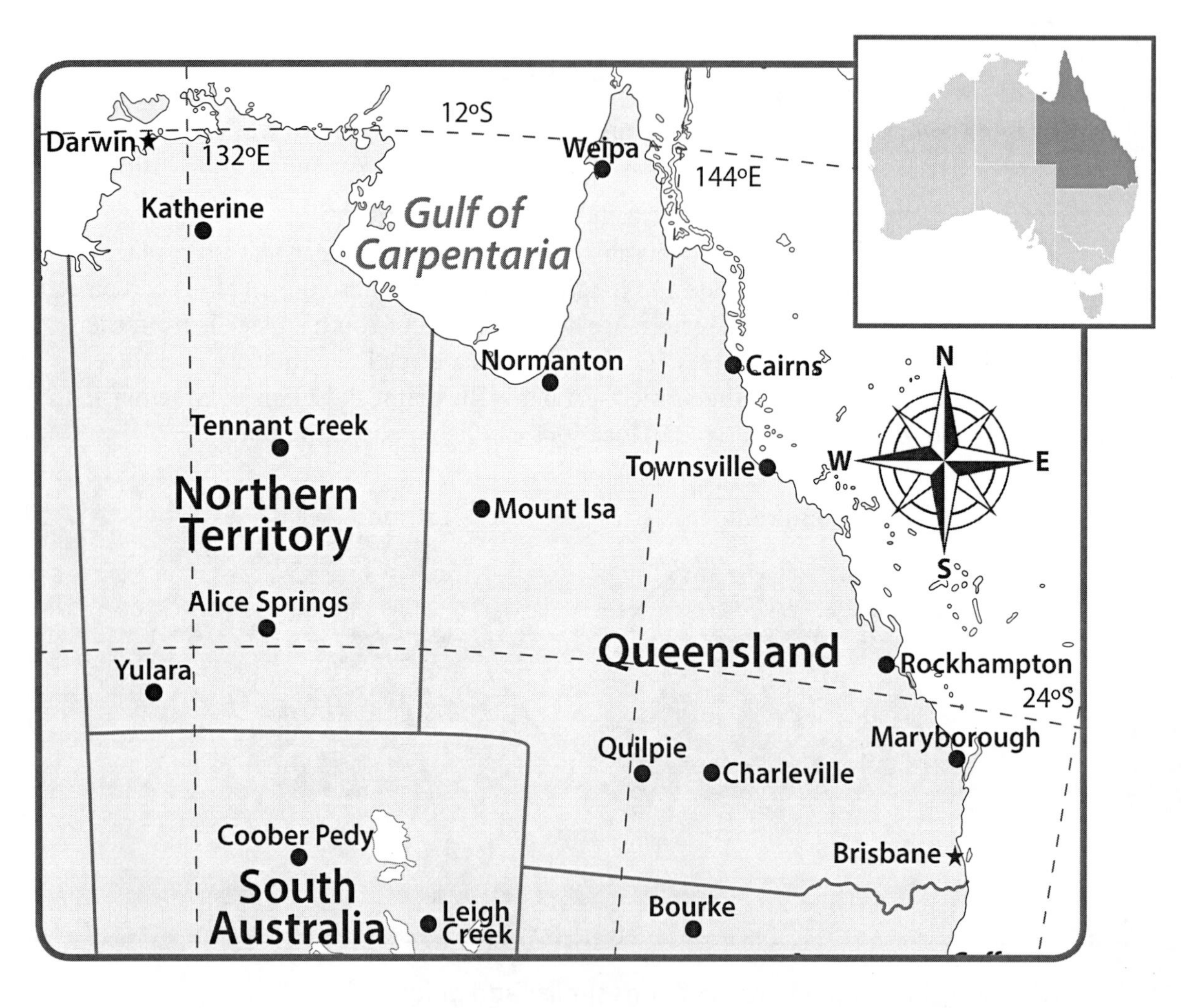

1. Choose one city in Queensland. List its relative and absolute locations.

   City: ______________________

   Relative Location: ______________________

   Absolute Location: ______________________

2. Why is it helpful to have two ways to give a city's location?

   ____

   ____

   ____

Name: ______________________________ **Date:** ______________

Directions: Think of a place in your community that you know well. Use other places to describe its relative location as accurately as you can. Give directions to this place from a different location. Then, draw a map of this place.

Place: ____

Relative Location

____

____

____

____

Directions

____

____

____

____

____

____

Name: ______________________________ Date: ______________

Directions: This is a map of Venezuela in South America. Use the map to answer the questions.

1. Which city is the capital?

____

2. Is Venezuela in the northern or southern part of South America? How can you tell from these maps?

____

____

____

3. Which city on this map is nearest Caracas? Which is farthest?

____

____

Name: ______________________________ **Date:** ______________

Directions: Use the information to complete the map of Venezuela's natural resources. Add the pictures from the legend in the appropriate locations.

Resources in Venezuela

(pickaxe)	mines	(drop)	oil	(diamond)	diamonds
(cow)	livestock	(cup)	coffee beans		

- Livestock is raised just east of Puerto Ayacucho.
- There are mines just south of Ciudad Guayana.
- Oil is found between Calabozo and Ciudad Guayana.
- Coffee beans are grown west of Caracas.
- Diamonds are found between Puerto Ayacucho and Santa Elena de Uairen.

Name: ______________________ Date: ____________

Directions: Read the text and study the photo. Then, answer the questions.

Venezuela's Oil Resources

Venezuela is a country a little larger than Texas. Yet, it has the second largest oil reserves in the world. There is a big demand for oil. Almost every person in the world uses it. Oil can be made into gasoline for cars, ships, and airplanes. It can be used to make electricity, which powers many devices. Oil is even used to make asphalt roads and plastic.

Oil wells and pipelines dot the northern shore of Venezuela, especially around the city of Maracaibo. But the largest oil reserves are found in the Orinoco Belt. It stretches from Calabozo to Ciudad Guayana. It has billions of gallons of crude oil.

Skilled workers drill through rock until they reach oil. Then, they use rigs to bring the oil up to the surface. Oil can be sold in its crude, or unrefined, form. Or, it can be refined and sold later.

Oil is a huge part of Venezuela's economy. It provides jobs for many people. And the country makes money from it when it is sold.

1. What can oil be used for?

____

____

____

2. Where is the largest oil reserve in Venezuela?

____

3. Why is oil important to Venezuela?

____

____

____

Name: ____________________ Date: ____________

Directions: Study the bar graph showing the amount of coffee beans Venezuela produced in different years. Then, answer the questions.

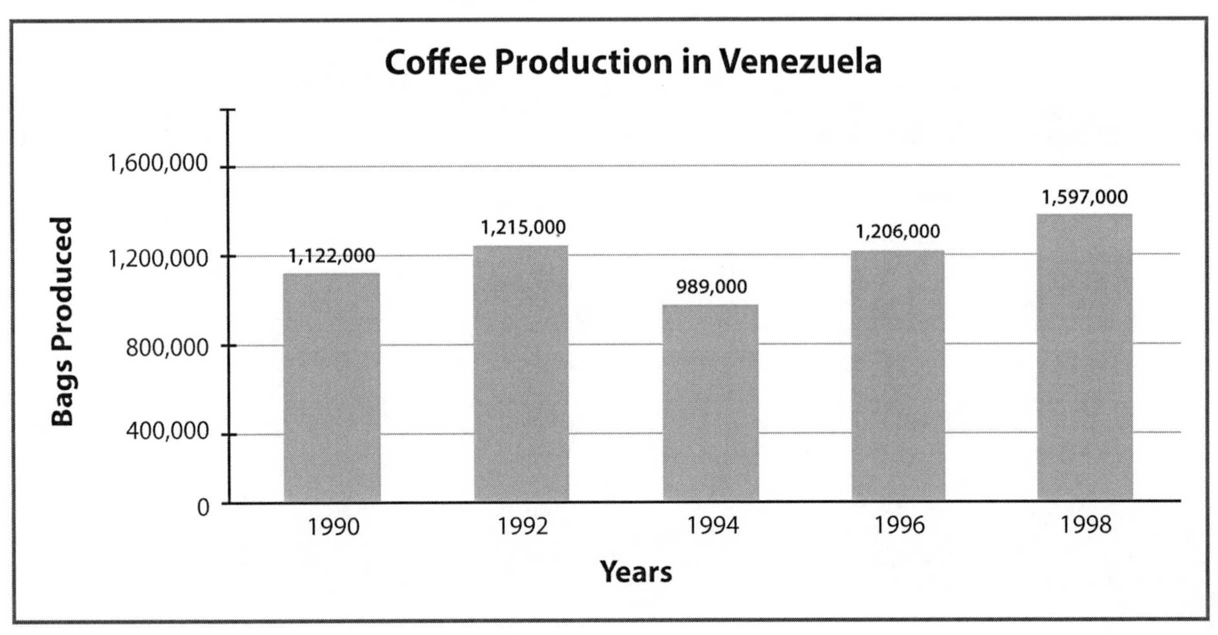

1. Describe the trend of this graph.

____

____

____

2. How might the weather contribute to a low year, such as 1994?

____

____

____

3. What do you notice about production between 1996 and 1998?

____

____

____

____

Name: ______________________________ **Date:** ____________________

Directions: Think of three ways you use oil every day. Then, imagine there was suddenly no more oil. How would you have to do things differently? Write and draw in the chart to show your ideas.

With Oil	Without Oil

Name: ______________________ Date: ______________

Directions: Study the map of Peru. Then, answer the questions.

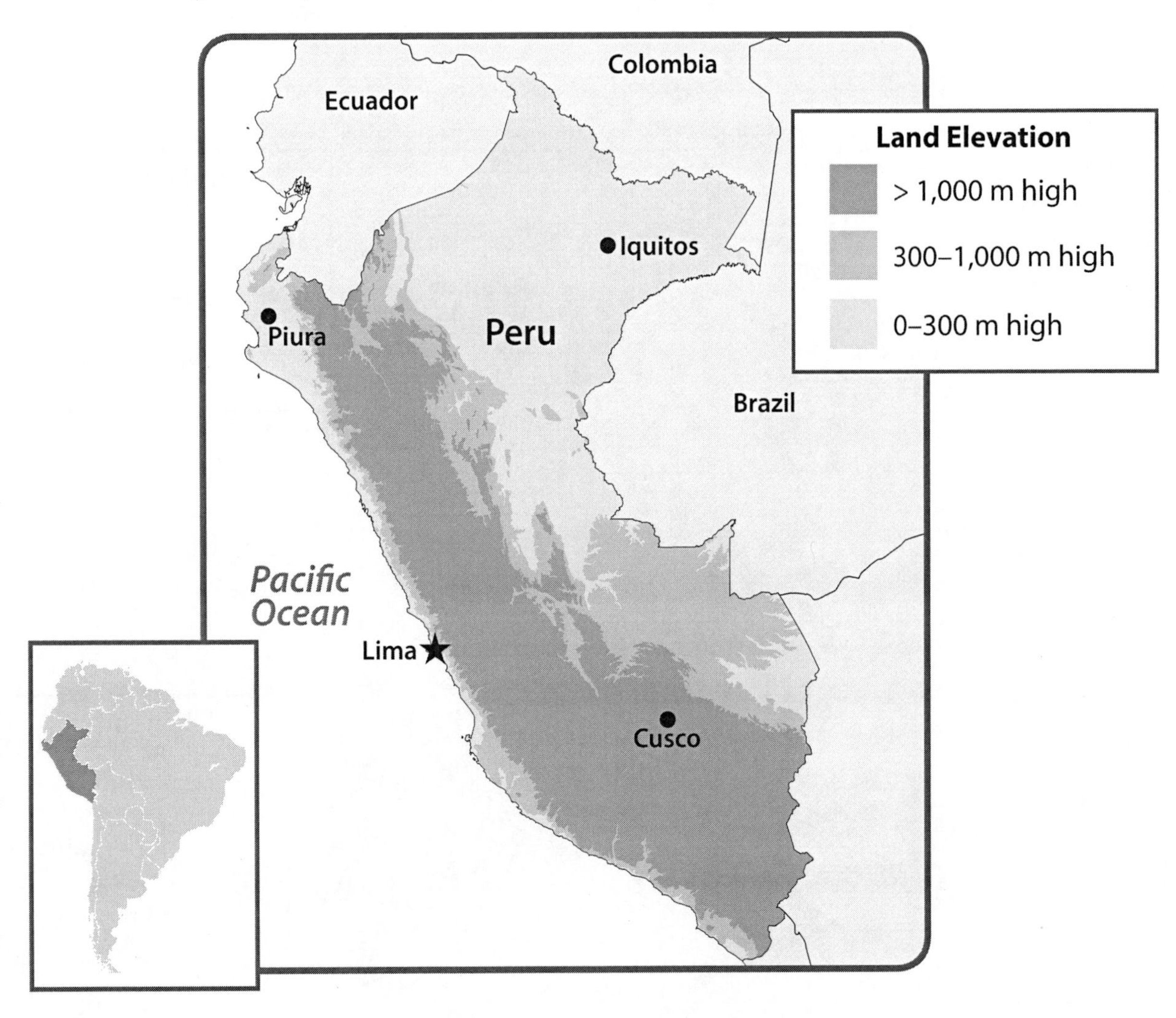

1. What is the elevation of Peru's capital city?

____

2. Which city shown has the highest elevation?

____

3. There are fewer cities at higher elevations. Why might people build fewer cities there?

____

____

____

Name: ______________________________ Date: ________________

Directions: Peru has three distinct regions with their own ecosystems. Follow the steps to complete the map.

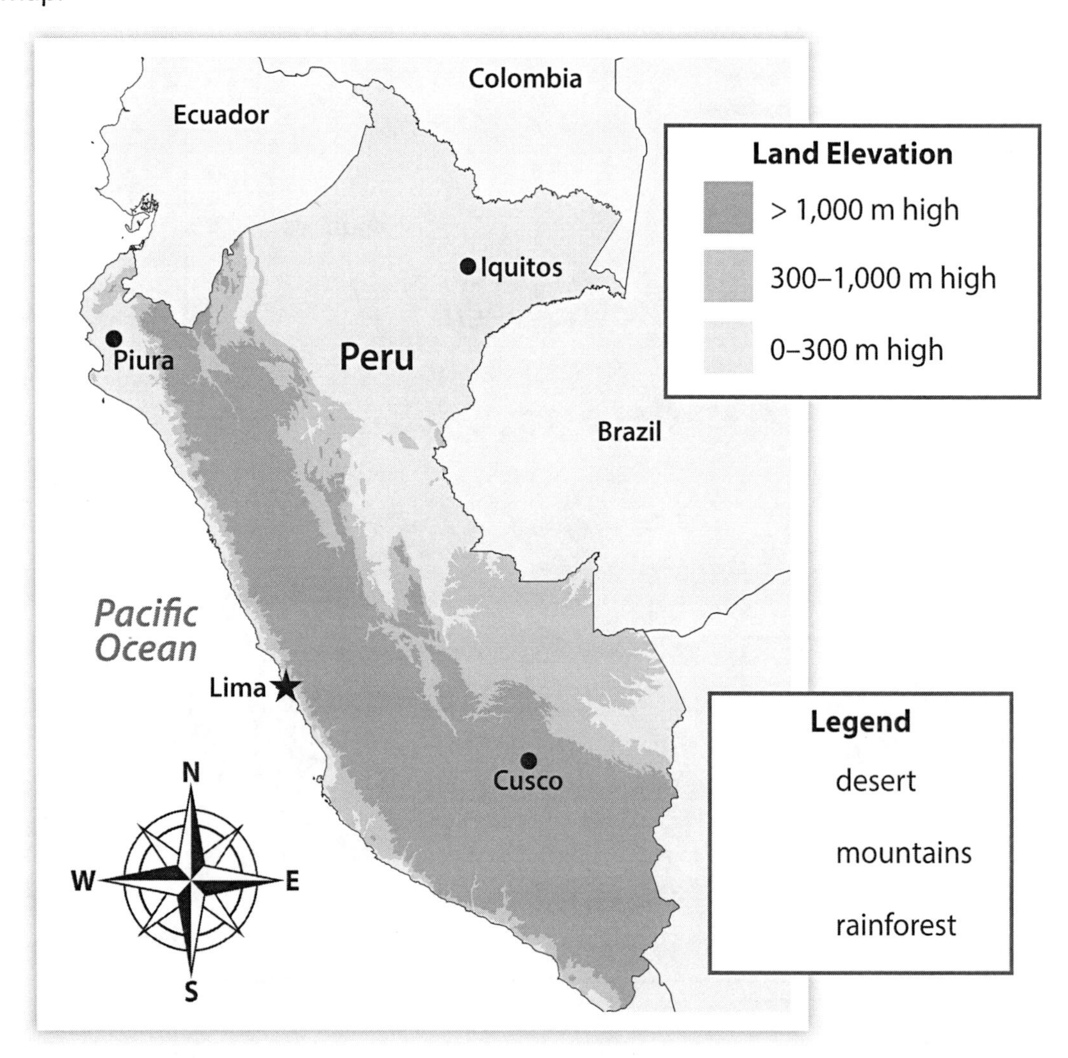

1. Create a symbol for each region in the legend.

2. The desert is a narrow strip along the west coast. Add your desert symbol to the map.

3. The mountains are just northeast of the desert, stretching from the northern to the southern parts of Peru. They are the higher elevations on the map. Add your mountains symbol to the map.

4. The rainforest takes up the northeastern part of the state. Add your rainforest symbol to the map.

5. Color the Pacific Ocean blue.

Name: ______________________________ **Date:** ______________

Directions: Read the text, and study the photos. Then, answer the questions.

Peru's Ecosystems

Peru has desert, mountain, and rainforest regions. Each region has different ecosystems. They also have different weather, temperatures, and animal life.

The desert on the west coast is dry. It can be hard to tell where the beach ends and the desert begins. Peru is home to one of the driest deserts in the world. Yet many types of plants and animals can be found there.

The Andes Mountains are in central Peru. The grasses that grow on the mountainside are food for animals, such as llamas and alpacas. Temperatures can be warm during the day and freezing cold at night.

The rainforest in Peru is part of the Amazon. Over 6,000 kinds of plants grow there. There are also many species of endangered animals. The weather is hot and humid. Few people live in this region, and much of the land is protected.

Peruvian desert

Peruvian mountains

Peruvian rainforest

1. Where is the desert located in Peru?

____

2. What do the mountains provide for llamas?

____

3. How might people live differently in each region?

____

____

____

____

Think About It

Name: ______________________ Date: ______________

Directions: Study the circle graph. Then, answer the questions.

Land in Peru

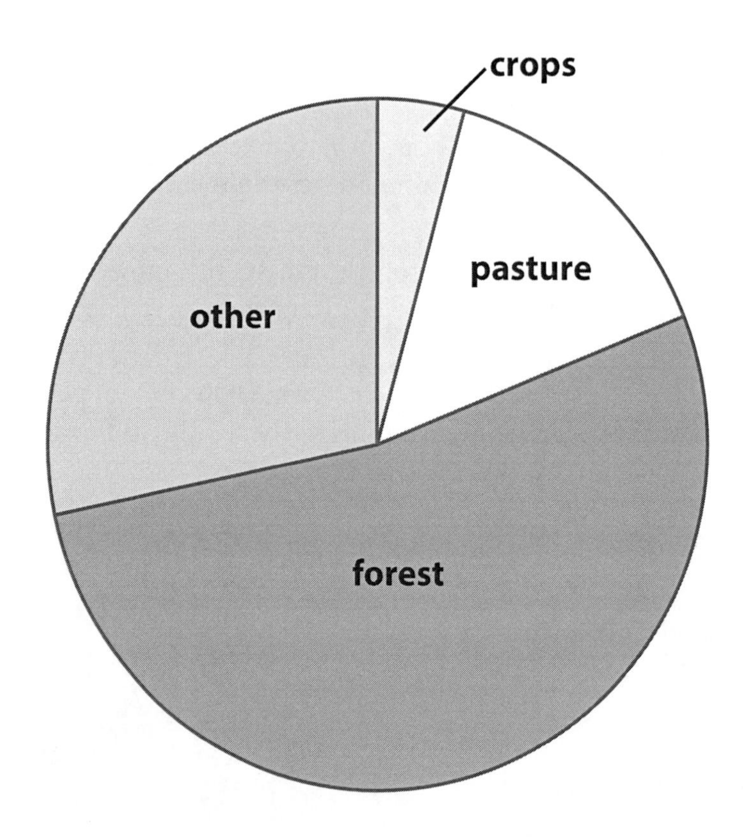

1. Would it be better for a farmer to raise crops or animals in Peru? Why?

____

____

____

2. What is most of the land in Peru used for?

____

3. What else do you think land is used for in Peru?

____

____

____

Name: ______________________ Date: ______________

Directions: Compare and contrast one region of Peru to where you live on the Venn diagram. Be sure to label the region you chose.

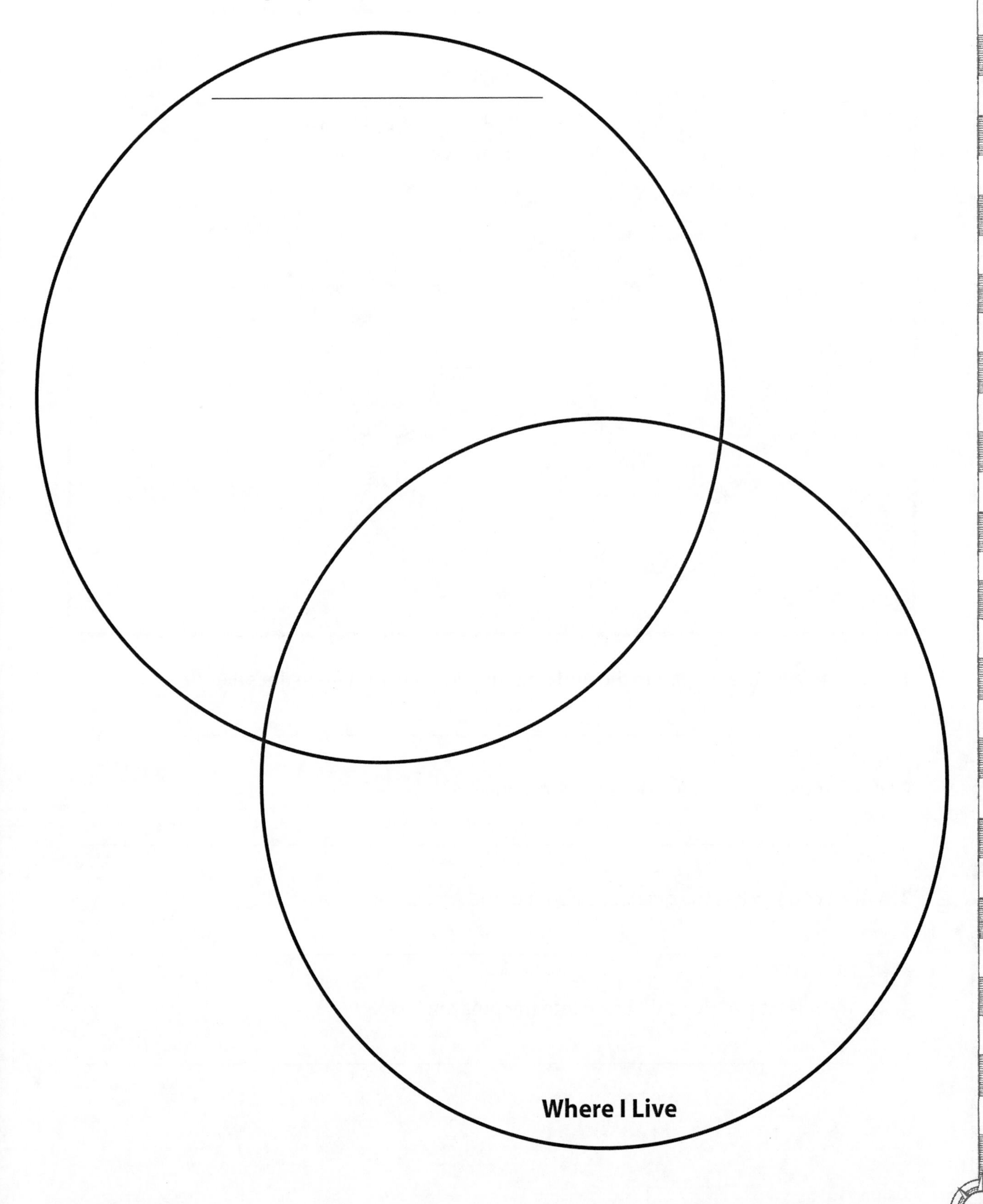

Name: ______________________________ Date: ______________

Directions: Study the map of the Amazon River in South America. Then, answer the questions.

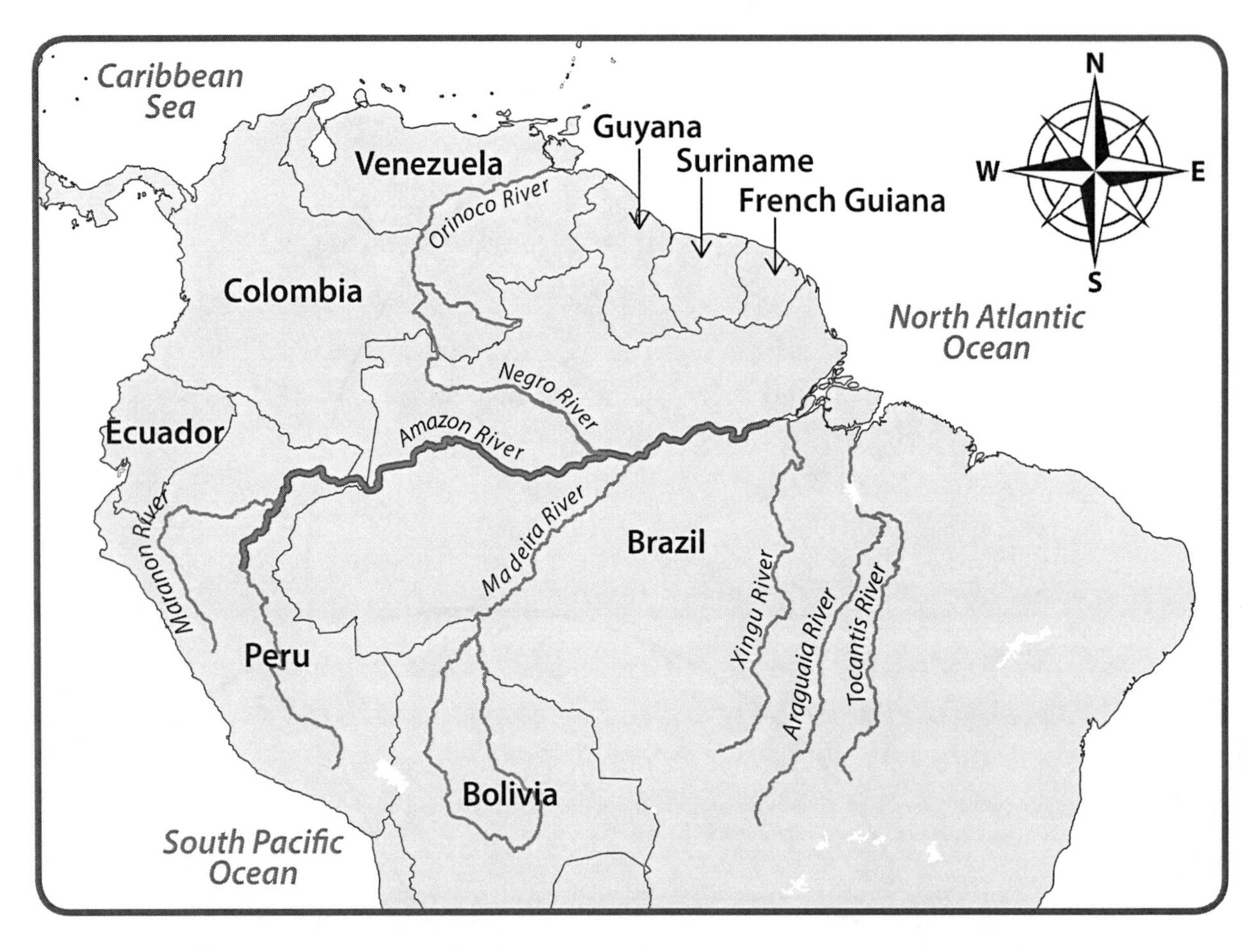

1. The Amazon River begins in the Andes Mountains of Peru. Where does it end?

____

2. In what general direction does the Amazon flow?

____

3. Which country has the greatest amount of the Amazon River in it?

____

4. Name at least two rivers that feed into the Amazon River.

____

Name: ______________________ Date: ______________

Directions: Follow the steps to draw a map of a small section of the Amazon River.

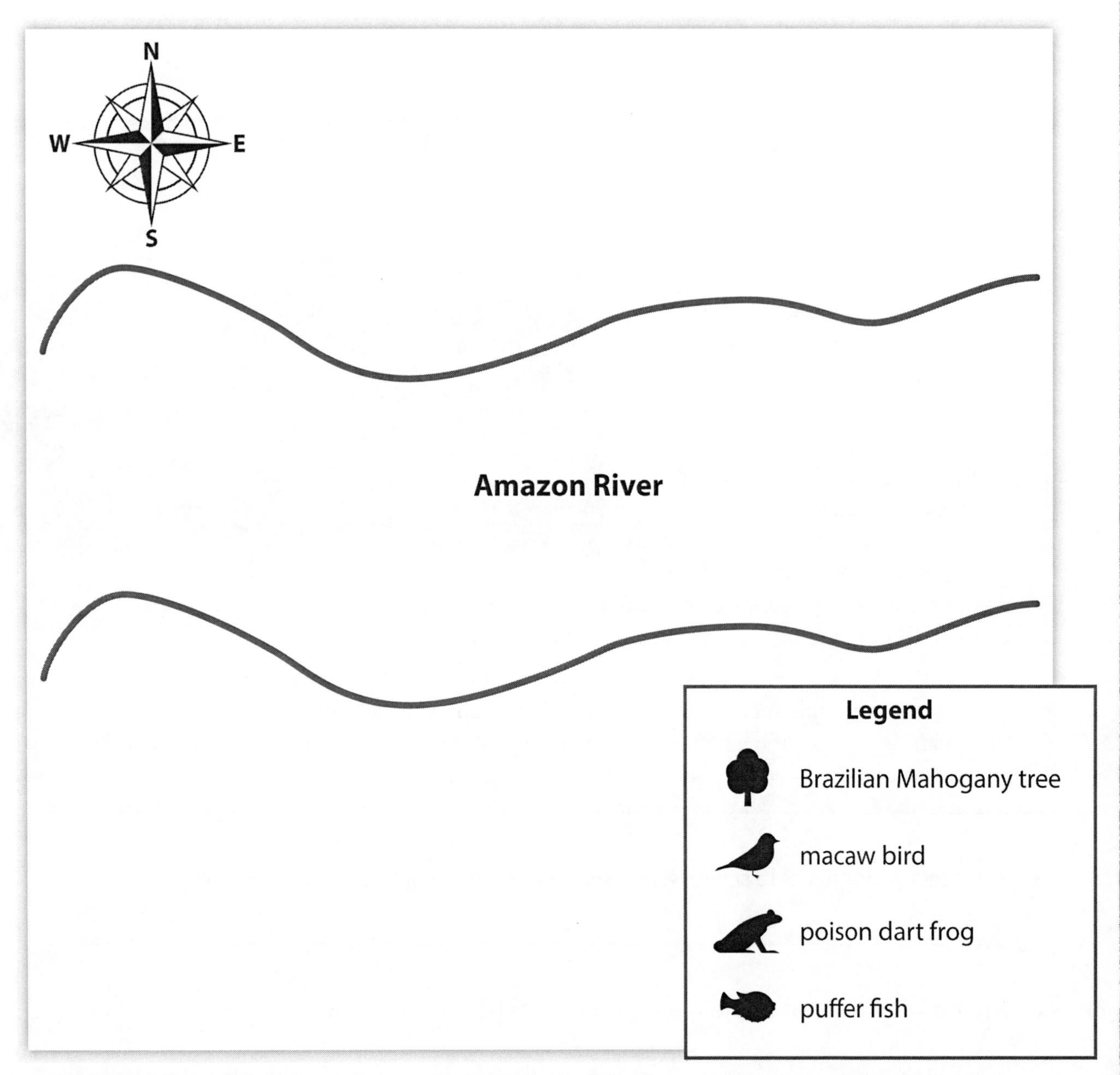

1. Draw a cluster of trees in the southwest corner of the map. Add more trees north of the river.

2. Draw a school of puffer fish heading east in the river.

3. Draw some frogs in the northeast corner of the map.

4. Draw some birds south of the river.

5. Color the river blue and the banks of the river green.

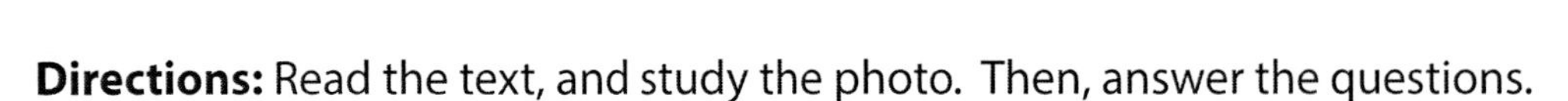

Name: ______________________ Date: ______________

Directions: Read the text, and study the photo. Then, answer the questions.

Secrets of the Amazon

The Amazon River is very big and very famous. Scientists have learned a lot about this river and the Amazon Rainforest. But, there is still much they do not know.

The Nile has often been called the world's longest river. But, the Amazon might be longer. The Amazon River begins somewhere in the Andes Mountains and ends at the Atlantic Ocean. But scientists are not sure exactly where it begins. The river's length was estimated at 4,000 miles (6,400 km). Recently, it was measured at 4,300 miles (6,900 km). GPS and satellite pictures help scientists measure it. But even those technologies are not perfect. The exact length of the Amazon River remains unknown.

Scientists know there are thousands of kinds of plants and animals in the Amazon. But, they still find new species. Some of the plants can be used as medicine. Doctors want to know if some of the plants can cure other sicknesses. Will we ever know all of the Amazon's secrets?

1. What technologies help scientists measure the length of the Amazon River?

____

2. What is known about where the Amazon River begins and ends?

____

____

____

3. Why do doctors care about plants found in the Amazon?

____

____

____

Name: ______________________ Date: ______________

Directions: Study the chart. Then, answer the questions.

Longest River on Each Continent		
River	**Continent**	**Length (in miles)**
Amazon River	South America	4,000
Missouri River	North America	2,540
Murray River	Australia	1,570
Nile River	Africa	4,132
Onyx River (glacial stream)	Antarctica	20
Volga River	Europe	2,193
Yangtze River	Asia	3,915

1. In the past, cities were often been built by big rivers such as these. Why?

____

____

____

2. Which river is the longest on the continent where you live?

____

3. How might the freezing cold weather in Antarctica explain why the Onyx River is so short?

____

____

____

4. Put the rivers in order from shortest to longest.

____

____

____

____

____

Name: ______________________________ Date: ________________

Directions: Some scientists think there might be people living in the Amazon who have never seen the outside world before. Imagine a group of these people was found. Write a letter to them explaining some of the most important things about your world.

Date: ____________________

____

____

____

____

____

____

____

____

____

____

____

____

____

____

____

Name: ______________________ Date: ______________

Directions: Study the map of Argentina. Then, answer the questions.

1. What countries border Argentina?

____

____

2. About how many miles long is Argentina from the northernmost point to the southernmost point?

____

3. About how many miles is it from Salta to Bahía Blanca?

____

4. About how many miles is it from Buenos Aires to Mendoza?

____

Name: ______________________________ **Date:** ____________________

Directions: Follow the steps to complete the map.

1. Draw a triangle 1,200 miles north of Río Gallegos.
2. Draw a star 300 miles east of San Juan.
3. Draw a square 600 miles northwest of Santa Fe.
4. Draw a circle 900 miles southwest of Bahía Blanca.
5. Outline Argentina in red.
6. Color the oceans blue.
7. Circle the capital of Argentina.

Name: ______________________ Date: ______________

Directions: Read the text, and study the photo. Then, answer the questions.

The Pampas

Argentina is a country with a lot of variety. In the north, it is closer to the equator, so the weather is tropical. The southern tip is so far from the equator that the weather is very cold. The country has mountains, forests, beaches, and plains.

The plains region is called Pampas. This is a very large area in the central part of Argentina. Starting at the Atlantic coast, it stretches west to the Andes Mountains. The western Pampas are very dry, but it is humid in the east. Buenos Aires and other cities in the eastern area are where most of the people of Argentina live.

The Pampas are very flat. Many different crops grow there, such as corn, grain, and alfalfa. People raise animals, such as sheep, alpacas, cattle, and horses.

1. How does Argentina have both tropical and very cold weather?

____

____

____

2. What are the Pampas?

____

3. How are the Pampas used by the population?

____

____

4. In which area of Argentina would you want to live? Why?

____

____

Think About It

Name: ______________________ **Date:** ______________

Directions: Study the photo. Then, answer the questions.

This photo shows a home near Buenos Aires, Argentina, in the early 1900s.

1. What clues in the photo show it is from the past?

____

____

____

____

2. Do you think the photograph was taken in the Pampas (plains), Andes Mountains, or the Atlantic coast?

____

3. Give two reasons to support your answer to number two.

____

____

____

____

Name: ______________________ Date: ____________

Directions: Study the circles. Each circle is labeled with a number of miles. List things in each circle that are about that far from your home.

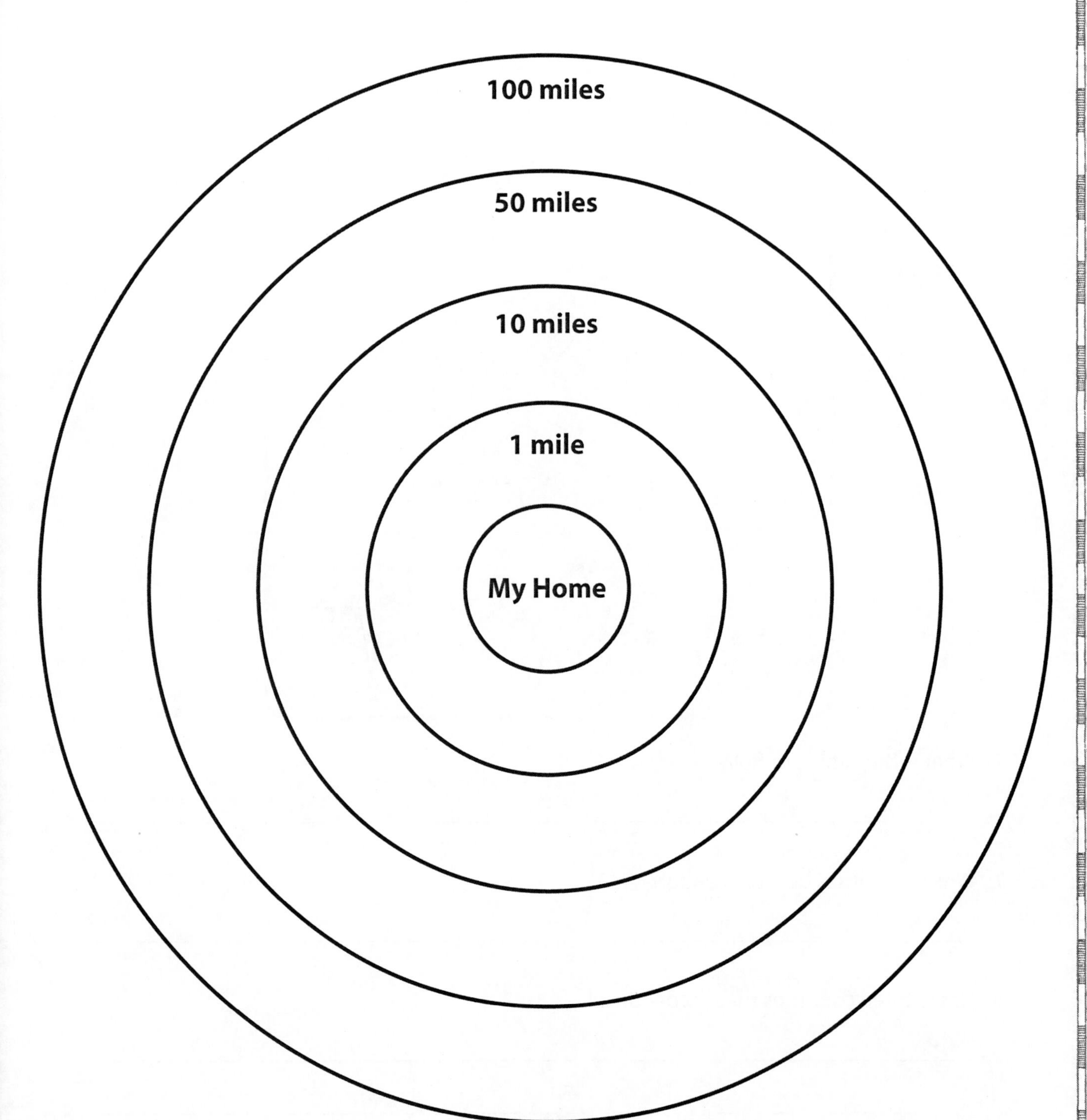

Name: ______________________________ **Date:** ________________

Directions: Study the map of South America. Then, answer the questions.

1. What is the capital of Bolivia?

____

2. Which countries border Ecuador?

____

3. Name three countries that border Brazil.

____

____

4. Which country lies west of Argentina?

____

Name: ______________________________ **Date:** ________________

Directions: Use the clues and the word bank to label the missing countries.

Word Bank		
Chile	Brazil	Peru
Colombia	Venezuela	Argentina

1. Brazil is the largest country in South America.
2. Chile is long and skinny.
3. Peru borders Bolivia and Brazil.
4. Colombia is north of Peru.
5. Argentina is east of Chile.
6. Venezuela is northeast of Colombia.

Read About It

Name: ______________________________ **Date:** __________________

Directions: Read the text, and study the photo. Then, answer the questions.

South American Transportation

Transportation in South America has changed a lot since World War II. The continent has always had many natural harbors. They have been updated over time, and most of the imports and exports are still carried by ship. Air travel has also increased over time.

Roads are the biggest and most important change in South America. Many people travel by car, and goods are shipped by truck. People are working hard to build roads that connect the countries. This will make travel between countries easier. There are challenges, though. South America has rainforests, and people do not want to cut down trees to make new roads. The Andes Mountains can also make road construction difficult. Some places far from big cities still have dirt or gravel roads. But easy travel is important, so South Americans will find a way.

1. How does South America transport most of its imports and exports?

____

2. Why are roads that connect countries important?

____

____

____

3. Why might the Andes Mountains pose a challenge for road construction?

____

____

____

____

Name: ______________________ Date: ______________

Directions: This chart shows about how many people there are for every car in five countries. Study the chart, and answer the questions.

Country	Number of People for Every Car
Argentina	3 people
Brazil	4 people
Colombia	14 people
Paraguay	19 people
Venezuela	7 people

1. Which country has the fewest number of cars? How do you know?

____

____

____

2. How do you think people travel in countries with few cars?

____

____

____

3. How could this information be used to decide where to build new roads?

____

____

____

____

4. How might car companies use this information?

____

____

____

____

Name: ______________________________ **Date:** ____________________

Directions: Draw and label four types of transportation you have used.

Name: ______________________ Date: ______________

Directions: This map shows diamond mines in South Africa. Study the map, and answer the questions.

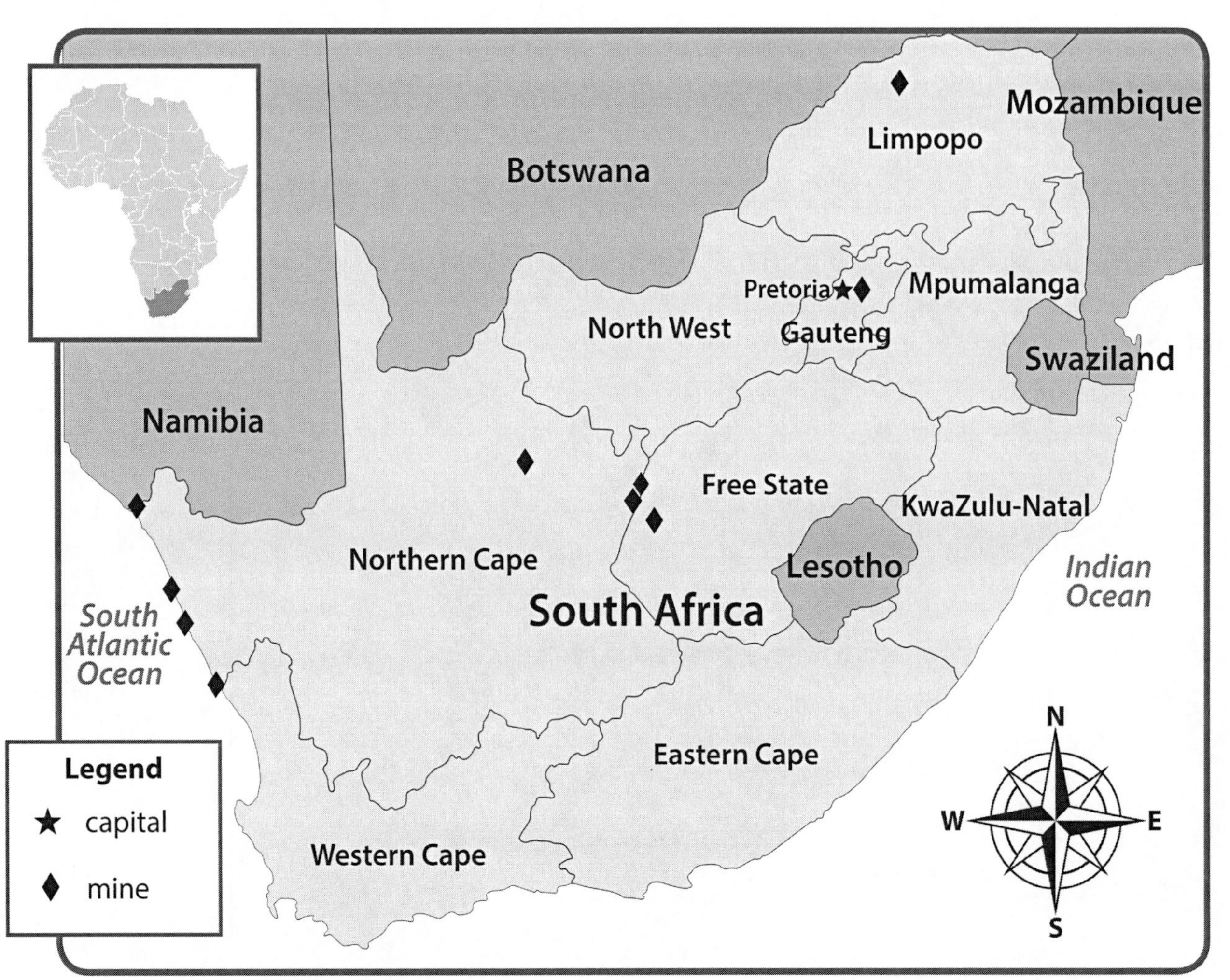

1. Which province has the most diamond mines?

____

2. Name two provinces that do not have diamond mines.

____

3. Which two countries are north of the Northern Cape province?

____

4. Describe where most diamond mines in South Africa can be found.

____

____

Creating Maps

Name: ______________________ Date: ______________

Directions: Follow the steps to create a treasure map.

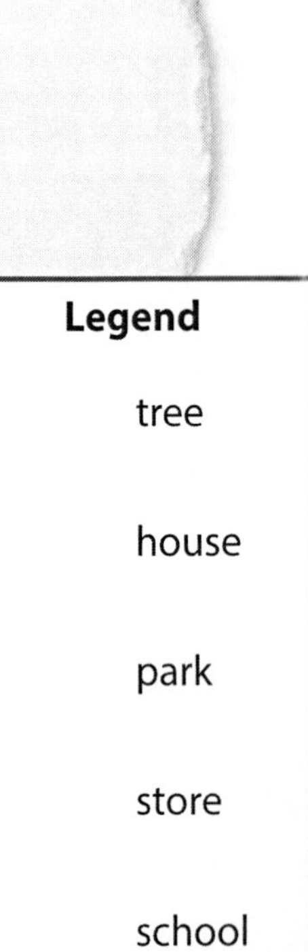

1. Draw a map of a town.
2. Mark your starting point with an *X*. Mark the hidden diamond with a diamond shape.
3. Add symbols to the legend.
4. Add the items from the legend to your map.
5. Write directions from the *X* to the diamond.

____

____

____

____

Name: ______________________________ **Date:** ________________

Directions: Read the text, and study the photo. Then, answer the questions.

Kimberley Diamonds

Diamonds are a natural resource. People use them in tools and in jewelry. Diamonds are found all over the world. Countries that have large supplies of diamonds can sell them for a profit. For over a century, Africa has been a leader in these jewels.

The first diamond in Africa was found in 1867. It was near Kimberley, South Africa. The gem was named Eureka. That means, "I have found it!" Almost overnight, people began mining for diamonds. They found many more! In only 20 years, Kimberley was producing 95 percent of the world's diamonds. The gems were found in other places in South Africa, too. Several mines popped up around the country.

South Africa was the leading producer of diamonds until the 1920s. But it still has mines open. They are not out of diamonds yet!

1. Where was the first diamond in Africa found?

____

2. Why do you think the first diamond was named *Eureka*?

____

____

____

3. What dangers could there be for the miners based on the photo?

____

____

____

____

WEEK 23
DAY
4

Think About It

Name: ______________________ **Date:** ______________

Directions: The chart shows how many diamonds were produced in different countries in 2015. Study the chart, and answer the questions.

2015 Diamond Production	
Country	**Millions of Carats**
Australia	10
Botswana	7
Congo	13
Russia	16
South Africa	1
Zimbabwe	4

1. How might diamond production help a country's economy?

____

____

2. Australia and Russia are much larger than the other countries. How could their size affect diamond production?

____

____

____

3. Which country produced the most carats of diamonds?

____

4. How many carats of diamonds were mined in these countries in 2015?

____

5. Which fact in this chart surprised you? Why?

____

____

____

Name: ______________________________ **Date:** ____________________

Directions: Not every country mines diamonds, but they do have natural resources. Make a list of at least 10 other natural resources. Put a star by any that you think are found in your area. Then, explain why those resources are important.

1. ____________________
2. ____________________
3. ____________________
4. ____________________
5. ____________________
6. ____________________
7. ____________________
8. ____________________
9. ____________________
10. ____________________

Reading Maps

Name: ______________________ Date: ______________

Directions: Study the map of Egypt. Then, answer the questions.

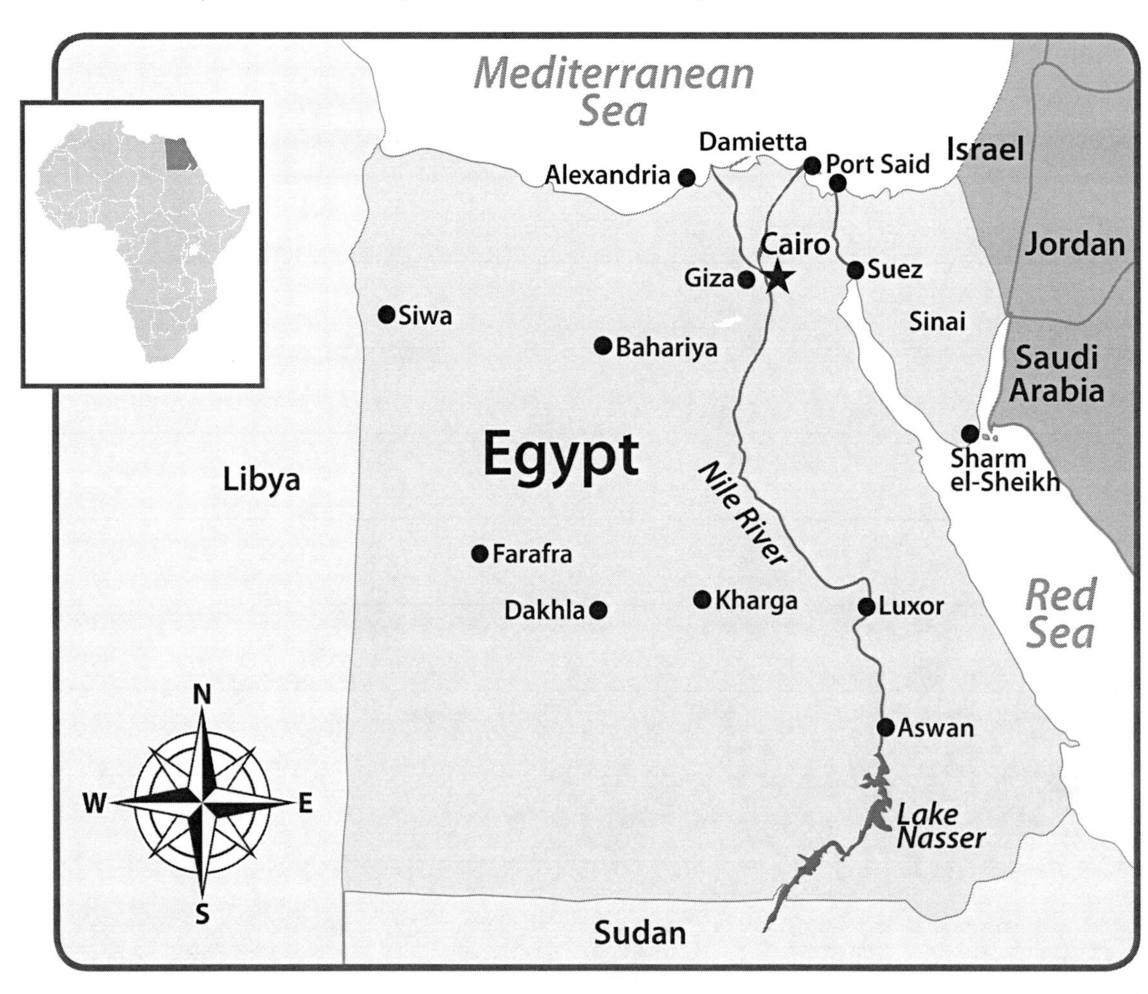

1. What is the name of the peninsula in northeastern Egypt?

____

2. Describe Giza's relative location.

____

____

____

3. What countries border Egypt?

____

____

Name: ______________________ Date: ______________

Directions: Use the clues to label the missing parts of the map.

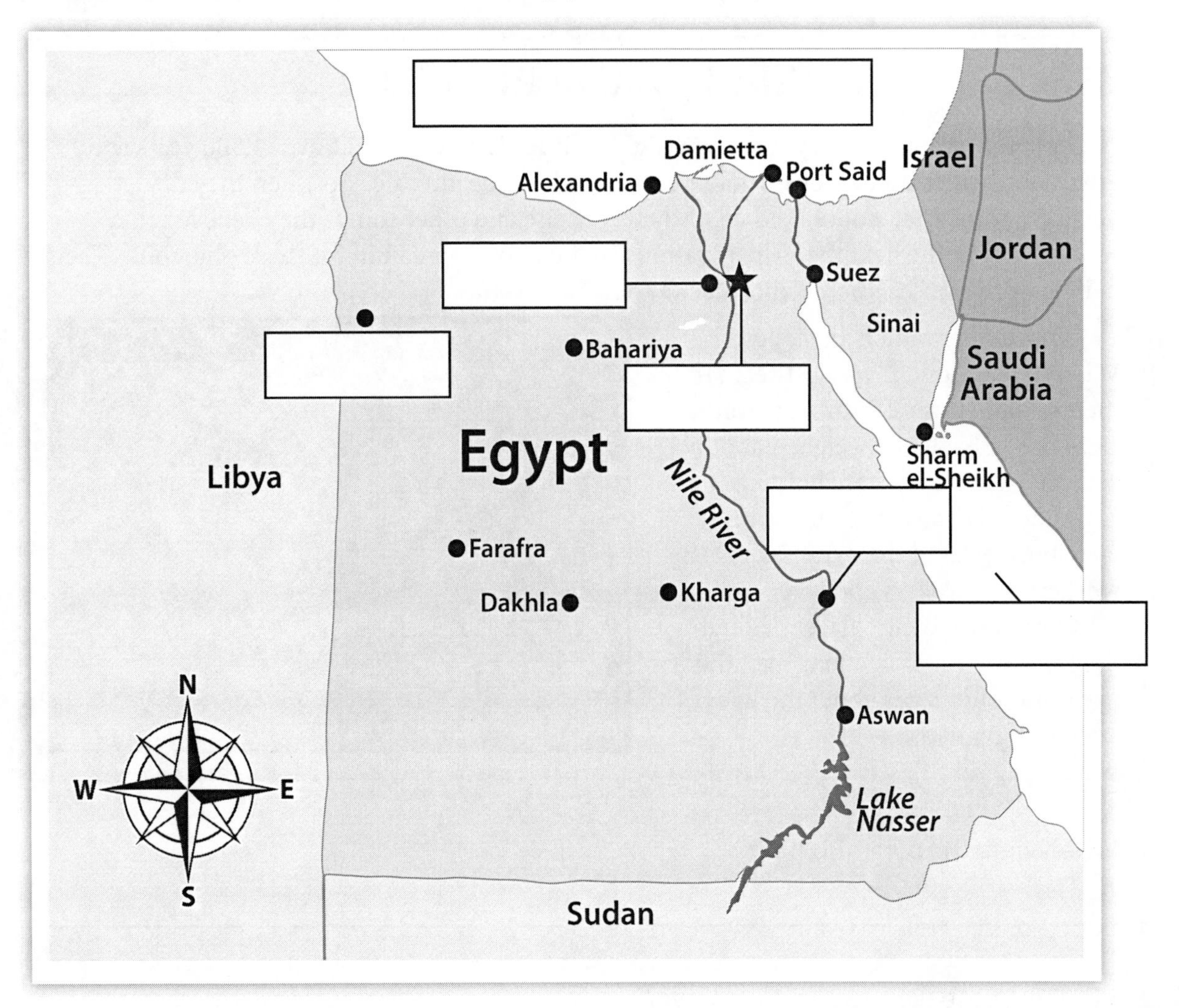

1. The Red Sea is east of the Nile River.
2. The Nile River empties into the Mediterranean Sea.
3. Cairo is the capital of Egypt.
4. Giza is just west of Cairo.
5. Luxor is also along the Nile River, but it is much farther south.
6. Siwa is farthest west on the map.

Read About It

Name: ______________________ Date: ______________

Directions: Read the text, and study the photo. Then, answer the questions.

The Egyptian Pyramids

Egypt is famous for its Great Pyramids of Giza. They were built over 4,500 years ago. Egyptians believed Pharaohs would become gods in the afterlife. So when they died, they needed a tomb that would hold all of their treasure and other things they believed they would need in the next life. The pyramids in the photo were built for three pharaohs. Their names were Khufu, Khafre, and Menkaure.

Khufu's pyramid is the tallest. When it was built, it was 481 feet (147 meters) tall. Over 2.3 million stones were used. Each one weighed about 2.5 tons. How they were built has been a mystery. Historians think ramps were built around the pyramid. Then, workers used rollers and levers to move each stone into place. Though the treasure was stolen long ago, the pyramids offer clues about the lives of ancient Egyptians.

1. When were the pyramids built?

____

2. What purpose did the pyramids serve?

____

____

____

3. How were the pyramids built?

____

____

4. For whom were these three pyramids built?

____

Name: ______________________________ Date: ______________

Directions: This graph compares the height of the Pyramid of Khufu to other landmarks. Study the graph, and answer the questions.

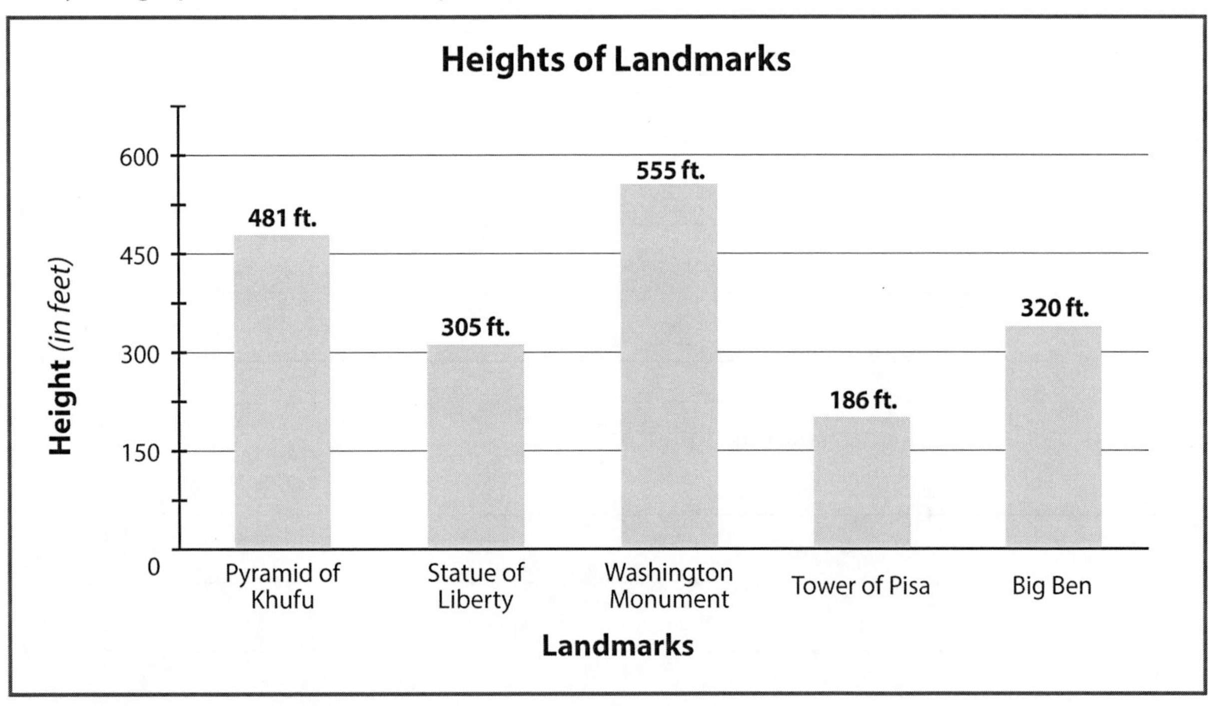

Think About It

1. What did you find most interesting or surprising about this graph?

____

____

____

2. The pyramid was built more than 3,500 years before any of the other monuments. What is impressive about the height of the pyramid?

____

____

____

3. Why might people want to build such tall monuments?

____

____

____

Name: ______________________________ **Date:** ________________

Directions: Think of an important landmark near you. Then, answer the questions.

Landmark: ____

1. Describe the landmark you chose.

____

____

____

2. Why is this landmark important or significant?

____

____

____

3. How might people remember this landmark in the distant future?

____

____

____

4. Draw a picture of the landmark.

Name: ______________________________ **Date:** ______________

Directions: Study the map of Zimbabwe. Then, answer the questions.

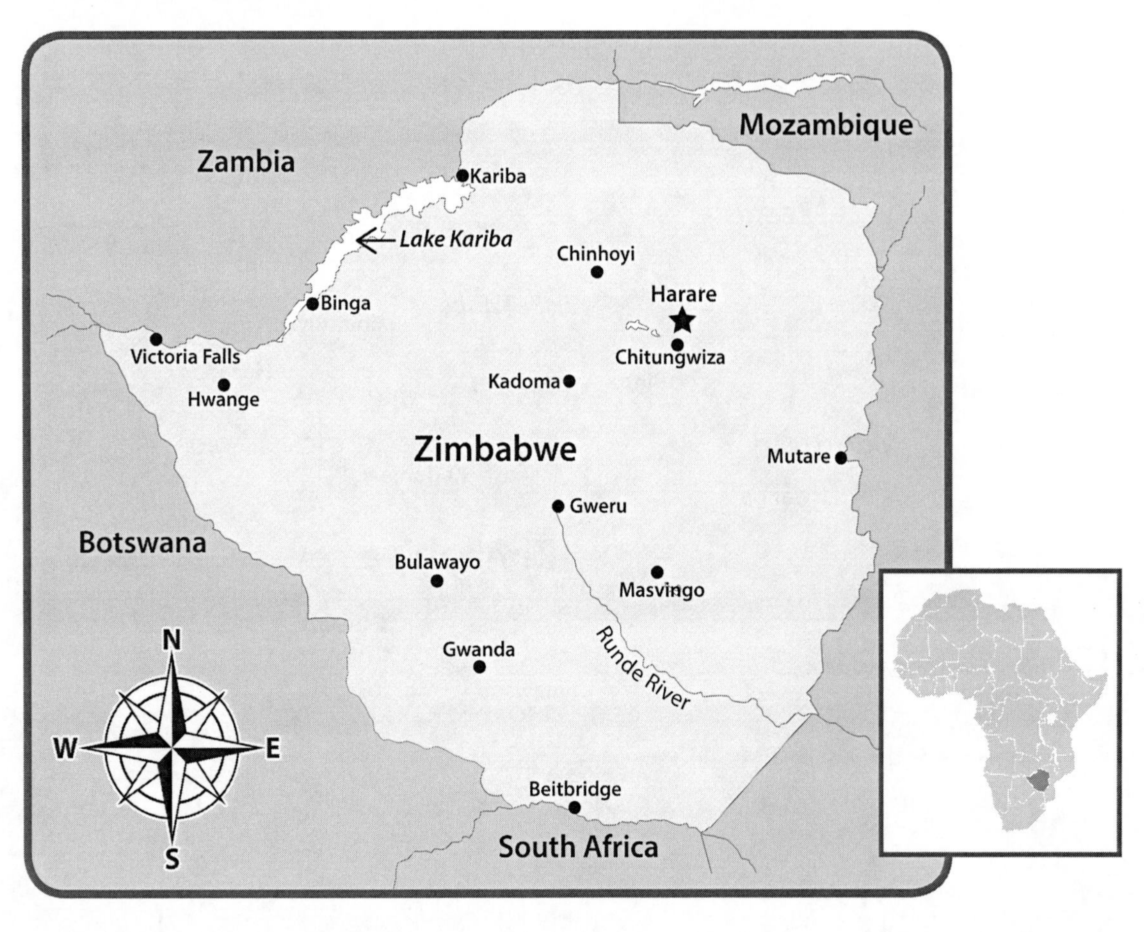

1. Describe the location of Zimbabwe in Africa.

____

____

2. How did you find the answer to question 1?

____

____

3. What is the easternmost city on the map?

____

4. What countries border Zimbabwe?

____

____

Creating Maps

Name: ______________________ Date: ______________

Directions: Follow the steps to make a language map for Zimbabwe.

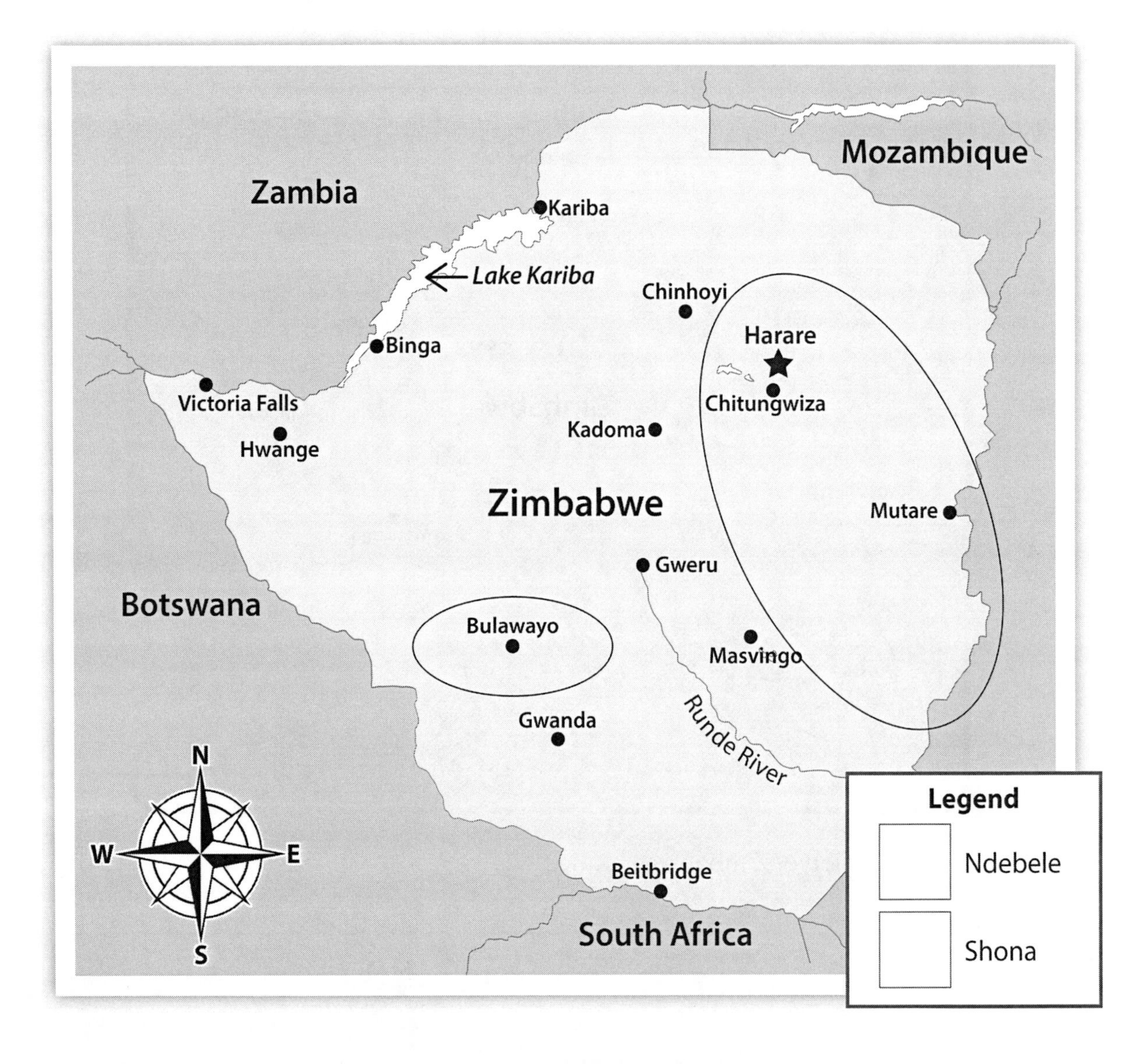

- The Ndebele language is spoken in the area surrounding Bulawayo.
- The Shona language is spoken by those living in the eastern half of Zimbabwe.

1. Use different colors to shade the languages in the legend.

2. Shade the map to show where these languages are spoken. Use the information in the box to help you.

3. Shade the countries that border Zimbabwe a different color.

Name: ______________________ **Date:** ______________

Directions: Read the text, and study the photo. Then, answer the questions.

Many Languages

Zimbabwe has 16 official national languages. This is more than any other country in the world. Zimbabwe was an English colony for many years. Many people can speak English because of this. Shona and Ndebele are two other popular languages. So, for a long time, these were the three official languages.

Speakers of other languages did not think this was fair. They were afraid the other languages would die out. So, leaders changed the constitution of Zimbabwe. They added 13 more languages to the official list. Under the new law, they all had to be promoted equally. This means all 16 languages have to be offered for public records and transactions. People hope this will help preserve their languages.

1. What were the first three official languages in Zimbabwe?

____

2. Why did people want to add other languages?

____

____

____

3. What might be the benefits and challenges of having 16 official languages?

____

____

____

____

____

____

Name: ______________________ Date: ______________

Directions: The circle graph shows the native languages of people in Zimbabwe. Study the graph, and read the text. Then, answer the questions.

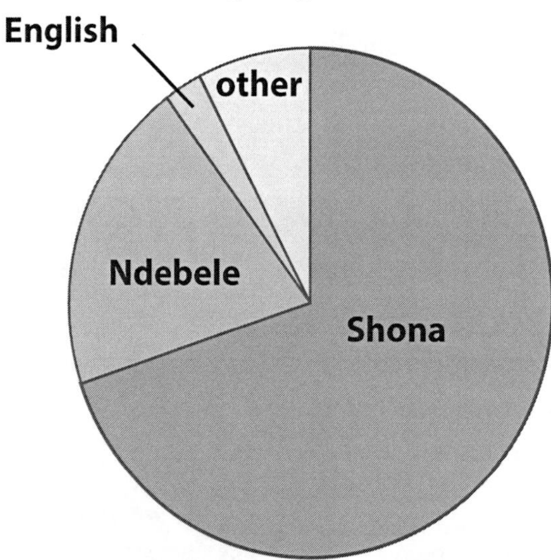

- Almost all people in Zimbabwe can speak English, but only the percentage in the graph consider it their native, or first, language.
- Other languages include: Chewa, Chibarwe, Kalanga, Koi-san, Nambya, Ndau, Shangani, sign language, Sotho, Tonga, Tswana, Venda, and Xhosa.

1. What does *native language* mean?

____

____

2. Why might people have been concerned that languages other than English, Shona, and Ndebele would die out?

____

____

3. Do you think Zimbabwe should keep all 16 languages as official languages? Why or why not?

____

____

____

Name: ______________________ Date: ______________

Directions: Answer the questions below.

1. What language(s) can you speak?

____

2. Besides your native language(s), what words or phrases do you know from other languages? Complete the chart to show this.

Language	Word or Phrase	Translation in English

3. The United States does not have an official national language. Do you think it should? Why or why not?

____

____

____

____

4. Describe the languages spoken in your community.

____

____

____

____

Reading Maps

Name: ______________________ **Date:** ______________

Directions: Study the map of Africa. Then, answer the questions.

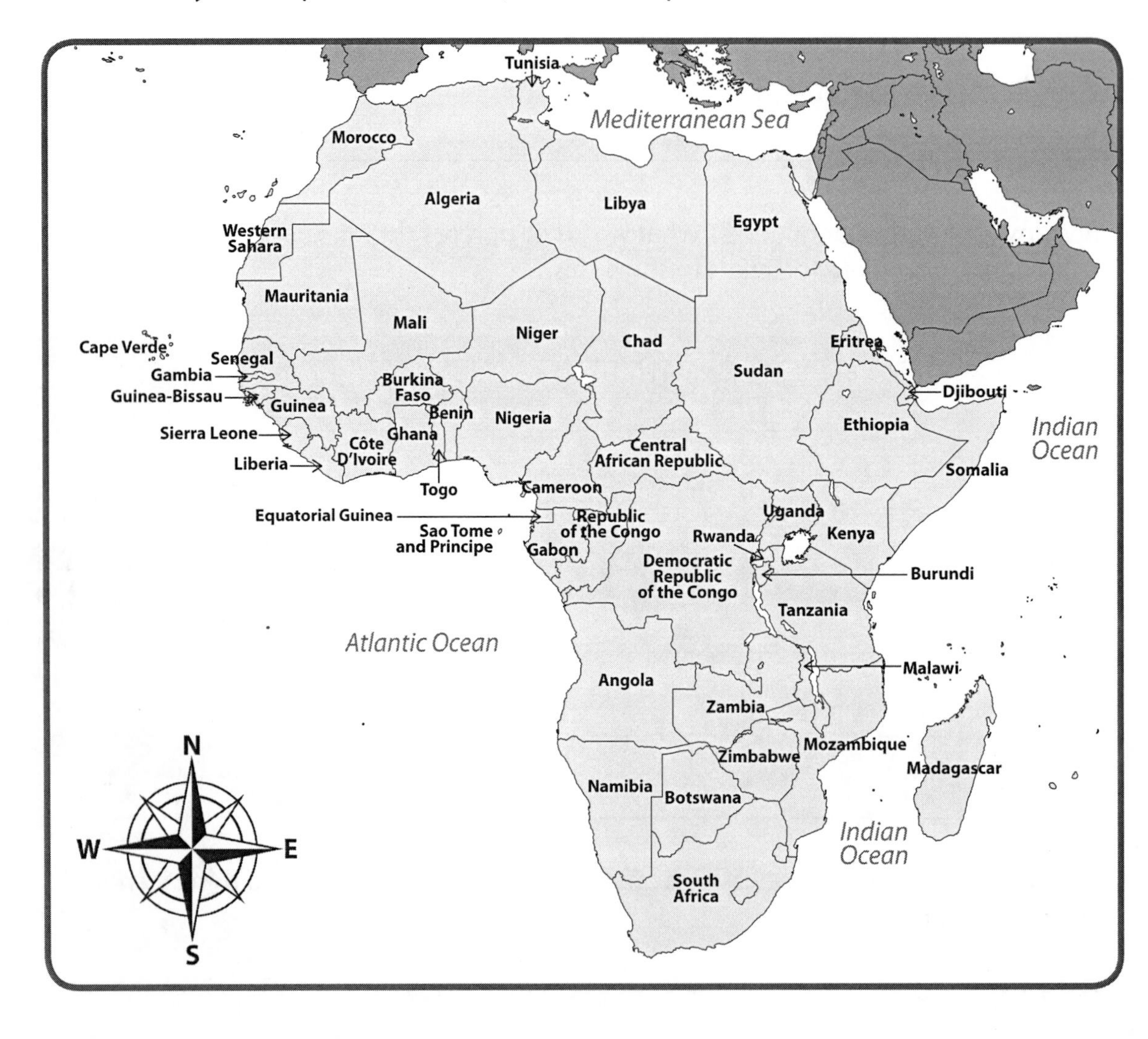

1. What country is north of Mozambique and south of Kenya?

____

2. Which country reaches the farthest north?

____

3. Describe Chad's location using each of the four cardinal directions.

____

____

____

____

Creating Maps

Name: ______________________________ **Date:** ____________________

Directions: This map shows a journey from Libya to Mauritania. Use cardinal directions to describe this trip through Africa.

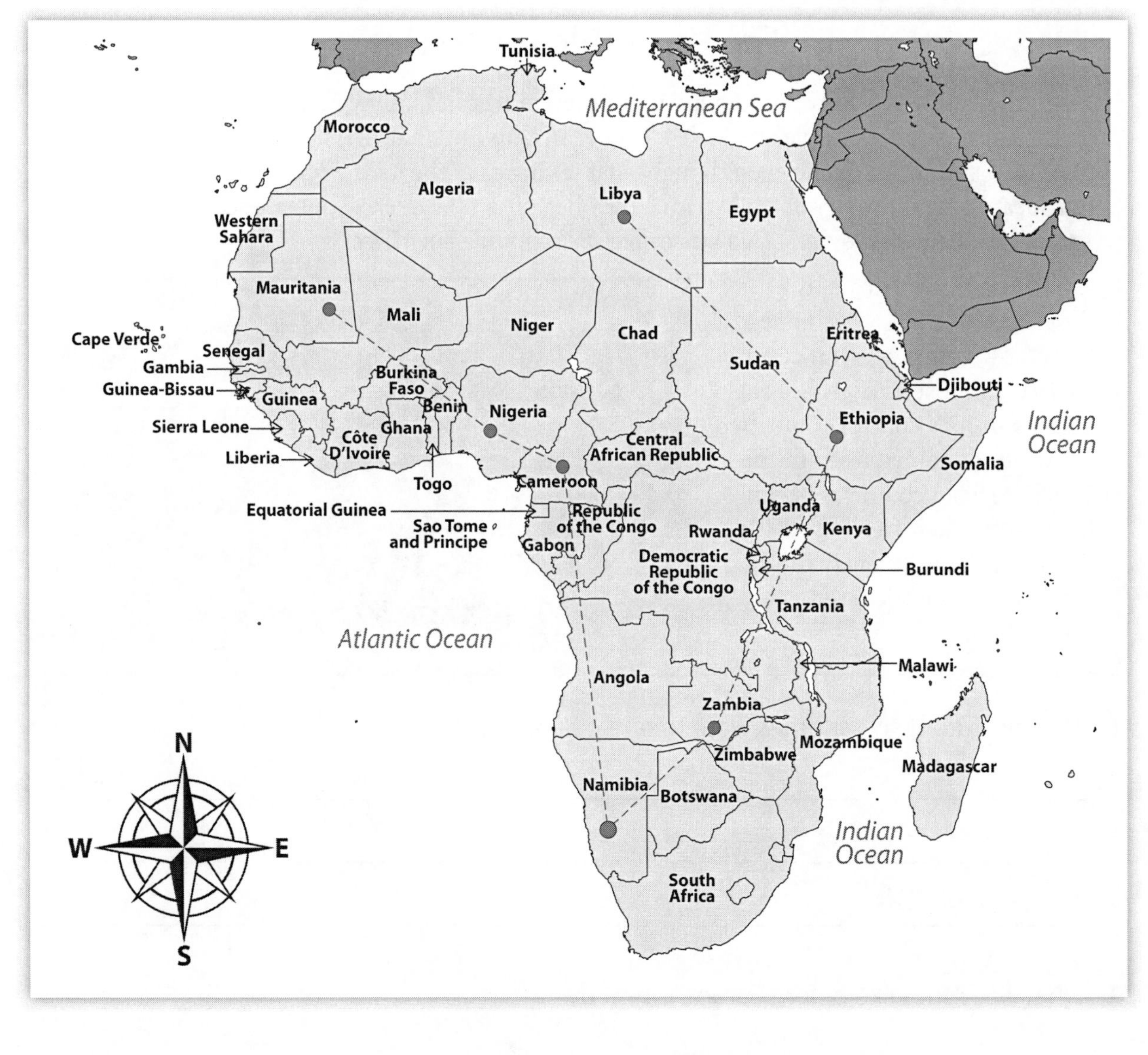

____

____

____

____

____

____

____

____

Read About It

Name: ______________________ **Date:** ____________

Directions: Read the text, and study the photo. Then, answer the questions.

The Scramble for Africa

In the late 1800s, European countries wanted more land. They had traded with people on the coast of Africa for years. So leaders began thinking that Africa could be good for their own countries. Some even thought that taking over the land would help Africans. European leaders sent troops to Africa. They began to take over. Countries competed for Africa's land and resources. This was called *the scramble for Africa.*

Before this, Africa had thousands of states. Each one had its own language and culture. By 1914, almost all of Africa was under Europe's control. Only two countries remained free. In the 1950s, African countries began fighting for their freedom. This lasted for the next few decades. Now, most countries rule themselves.

1. Why did European countries want to take over Africa?

____

____

____

2. What was Africa like before Europe took it over?

____

____

____

3. What effect did European countries have on Africans?

____

____

____

Name: ______________________ **Date:** ______________

Directions: Study the maps, and answer the questions.

Africa in 1914

Africa Today

Think About It

1. Name at least two differences you see between these two maps.

____

____

____

____

____

____

____

____

____

____

2. How do you think the lives of Africans have changed since 1914?

____

____

____

____

____

____

____

____

____

____

Name: ______________________________ Date: ______________

Directions: Use relative locations to write directions from one place to another. Use the ideas in the box, or think of your own locations. Then, draw a map to show your directions.

- your home to a friend's home
- your home to school
- your classroom to the office
- your bedroom to the kitchen

Directions from ______________________ to ______________________

____

____

____

____

____

____

____

Name: ______________________________ **Date:** ______________

Directions: Study this map of Africa. Then, answer the questions.

Saharan and Sub-Saharan Africa

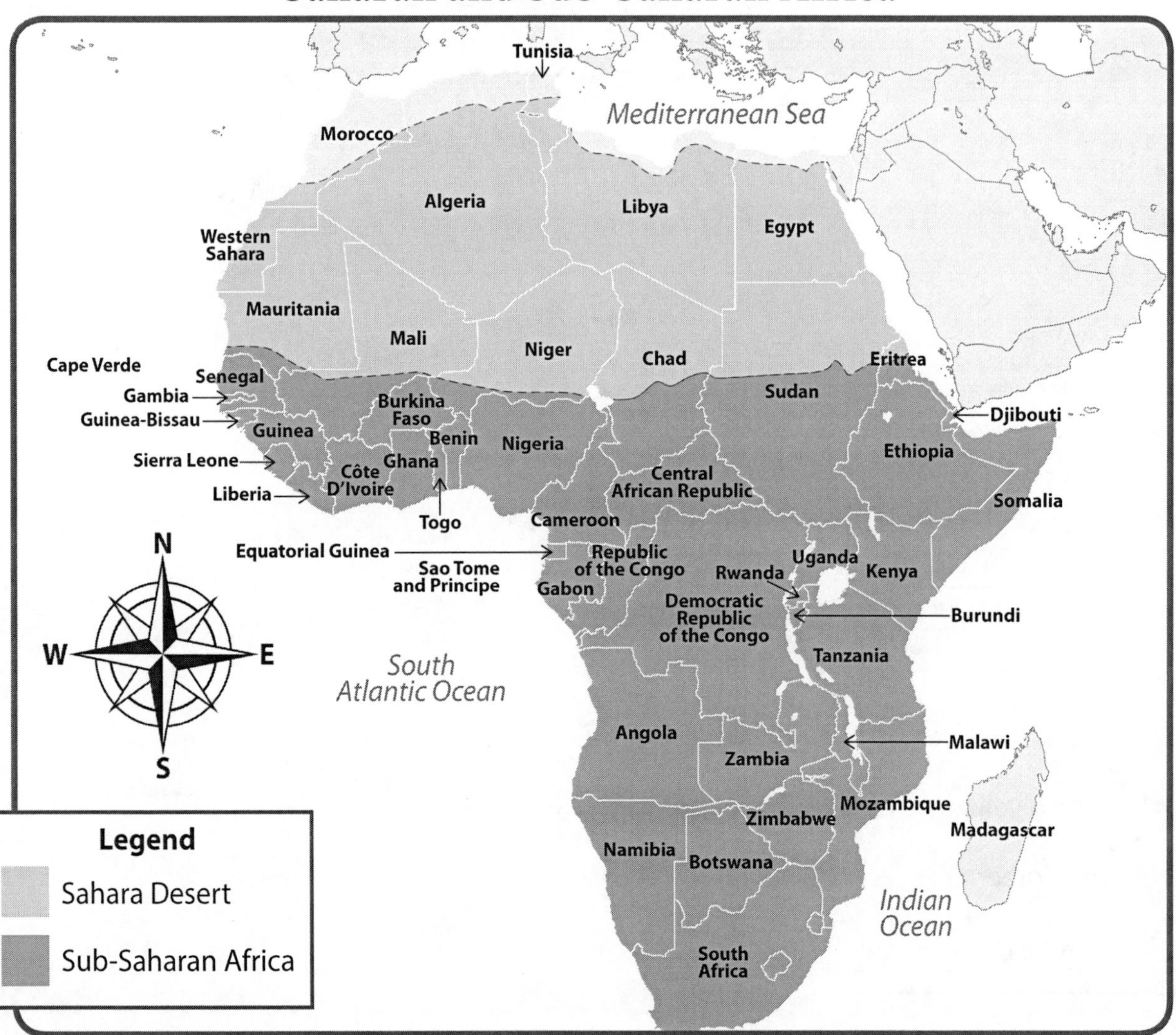

1. Describe the location of the Sahara Desert.

____

____

____

2. What do you think the climate is like in the Sahara Desert countries? Why?

____

____

____

Name: ______________________________ **Date:** ______________

Directions: Complete the map to show some of the climate differences in Africa. Color the countries listed in the chart using the color code in the legend.

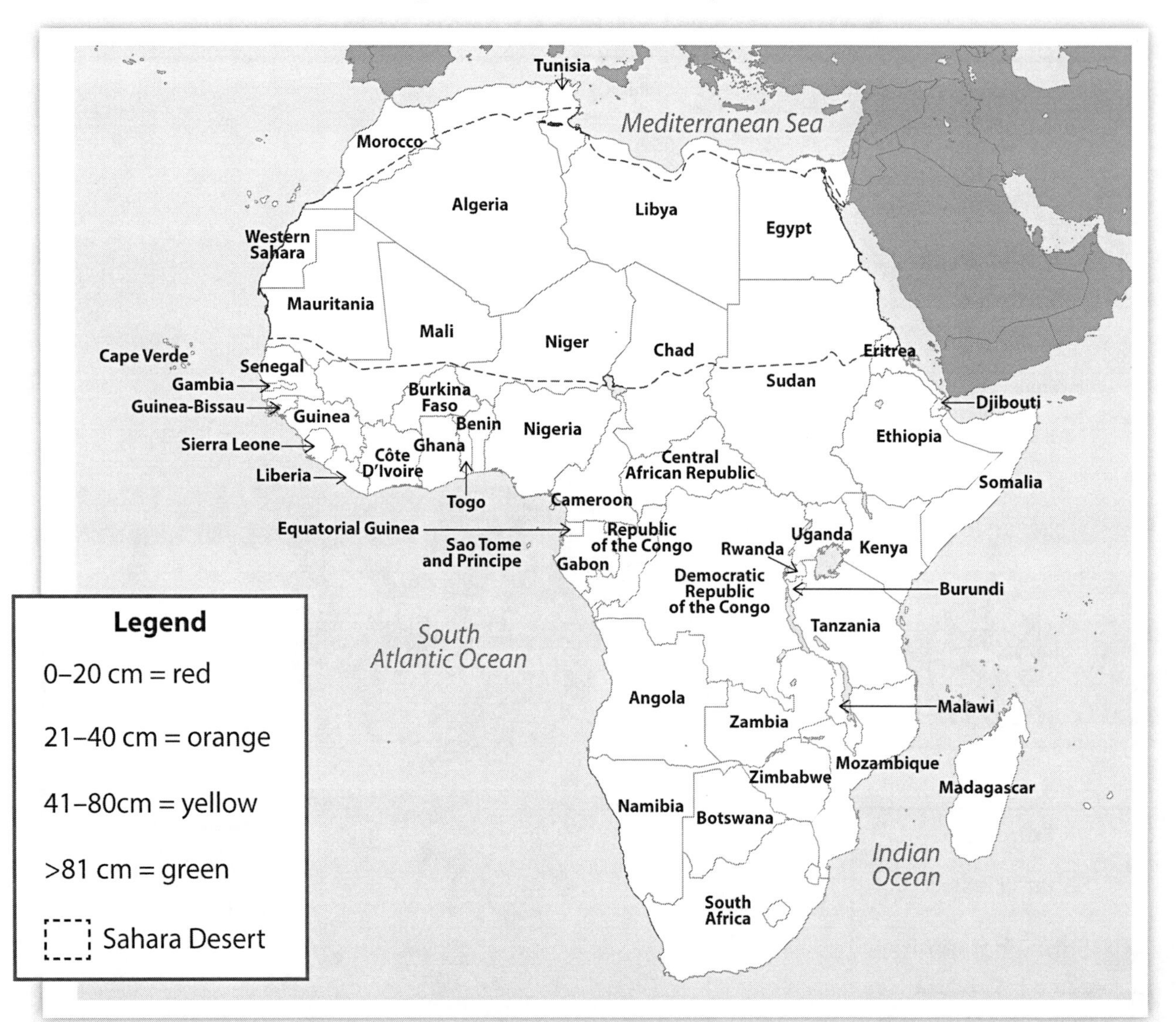

Average Precipitation					
Country	**Rainfall (in cm)**	**Country**	**Rainfall (in cm)**	**Country**	**Rainfall (in cm)**
Algeria	8.9	Egypt	5.1	Niger	15.1
Angola	101.0	Ethiopia	84.8	South Africa	49.5
Botswana	41.6	Libya	5.6	Sudan	25.0
Cameroon	160.4	Mali	28.2	Tanzania	107.1
Chad	32.2	Mauritania	9.2	Zambia	102.0
Congo	154.3	Namibia	28.5	Zimbabwe	65.7

Name: ______________________ Date: ____________

Directions: Read the text, and study the photo. Then, answer the questions.

Africa Divided

The Sahara Desert divides Africa into two climates. In some ways, it also divides the people. The Saharan countries are also called North Africa. In the past, the people living there were separated from the rest of Africa. Those in the Sub-Saharan did not travel north. The desert was just too harsh.

North African countries are close to the Middle East. These countries traded with the Arab nations. Travel between the two regions was easier. Over time, North Africans came to identify with the Middle East. The main religion of both is Islam. North Africa is also more developed like the Middle East.

Sub-Saharan Africa was different. It was made of many different groups. Each group had its own language, religion, and culture. Groups traded with each other, but kept their own identities.

Now, travel in Africa is easier. But in some ways, the divide still exists.

1. Why were the people in Saharan and Sub-Saharan Africa divided?

____

____

____

2. In what ways does North Africa identify with the Middle East?

____

____

____

3. How are the two regions of Africa different?

____

____

____

Name: ______________________________ Date: ____________

Directions: Study the pictures showing different climates in Africa. Then, answer the questions.

rainforest

desert

grassland

1. What three types of climates are shown?

____

____

2. How might life be different for people living in each type of climate?

____

____

____

3. Which climate would you most like to visit? Why?

____

____

Name: ______________________________ Date: ______________

Directions: Complete the web to describe the climate where you live. You may use the words in the box to help you.

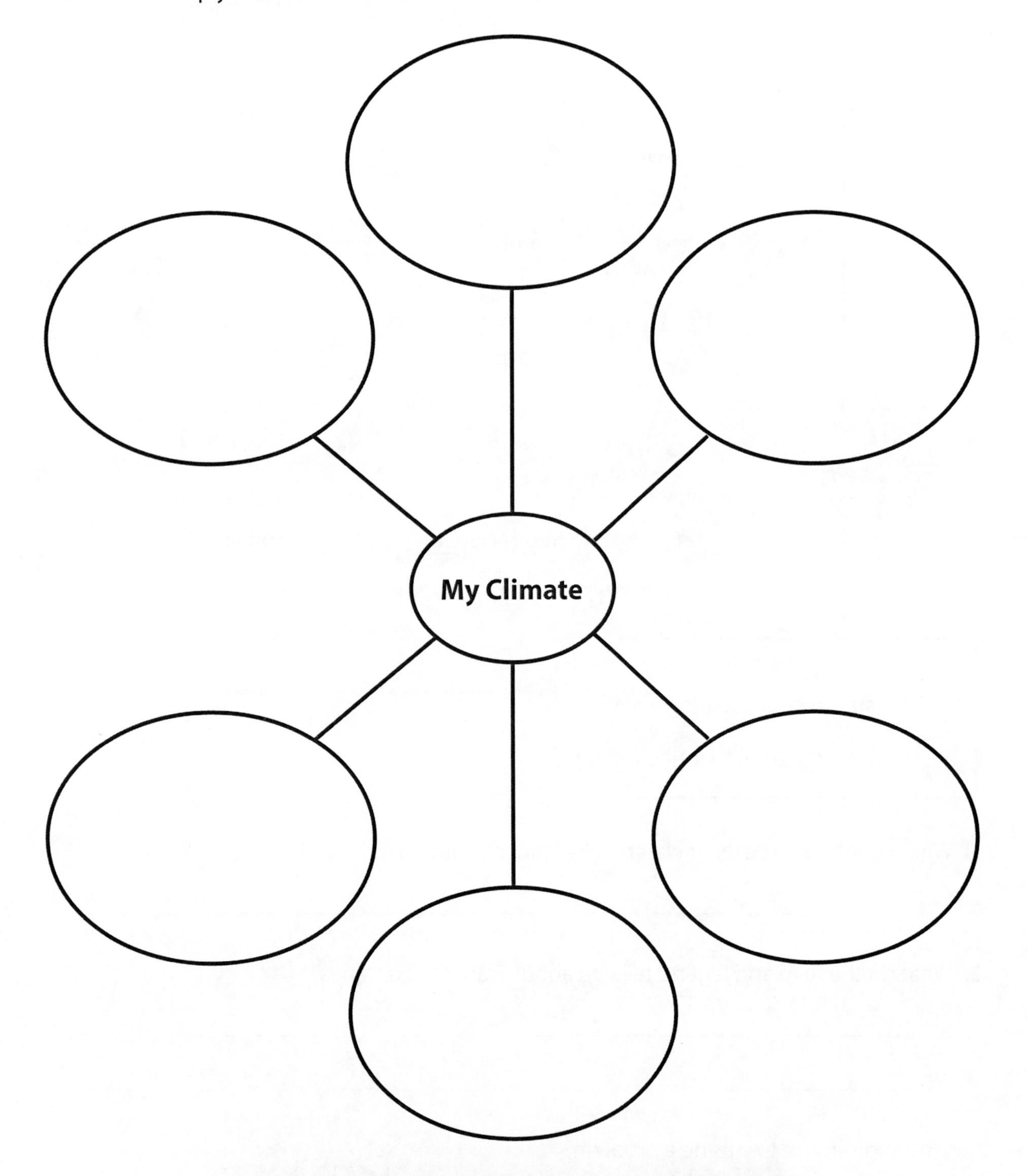

Climate and Weather Words		
hot summer	mild summer	cold winter
mild winter	rainy	dry

Name: ______________________ **Date:** ______________

Directions: Study the map of France. Then, answer the questions.

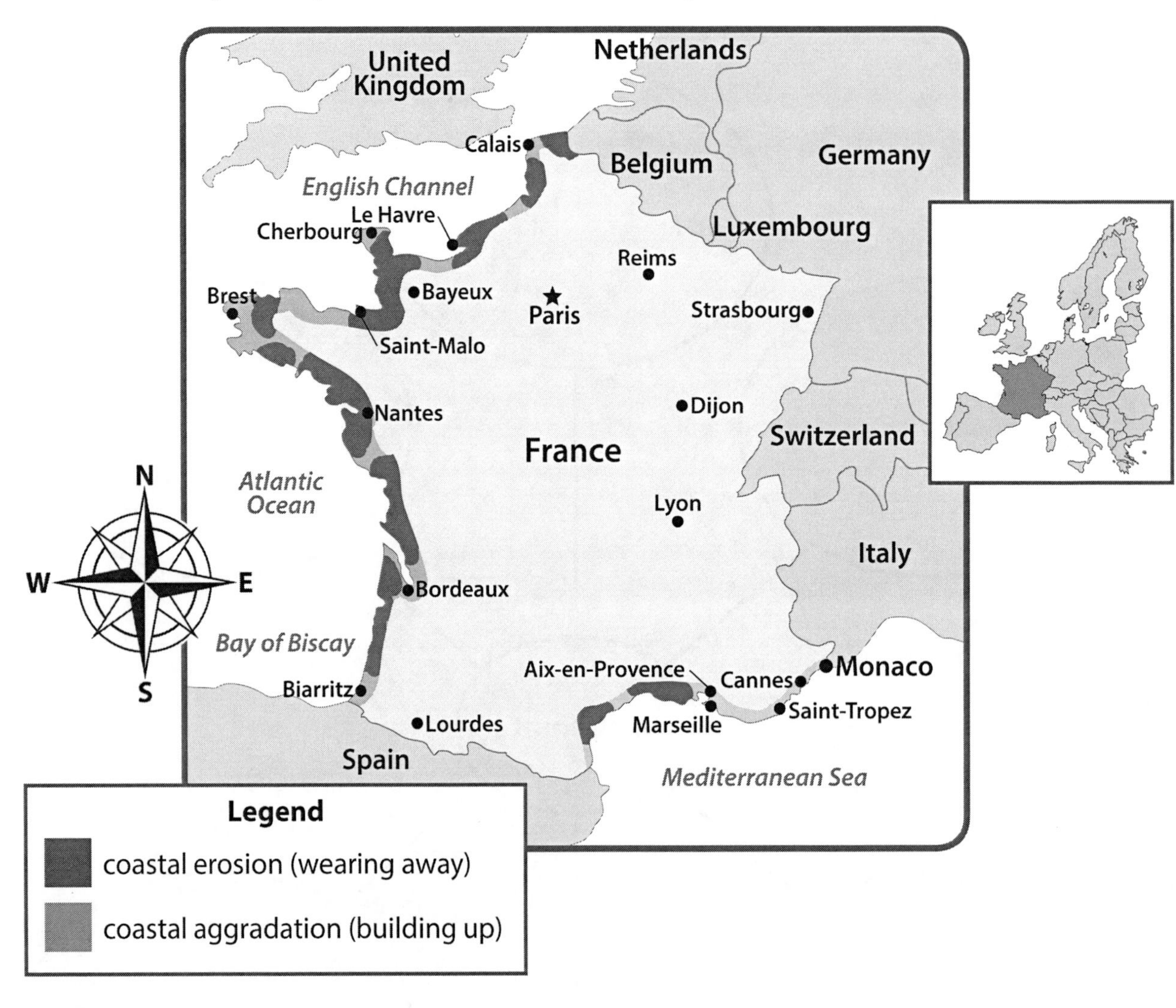

1. What country and body of water make France's southern border?

____

2. What do the map and legend tell you about France's coastlines?

____

____

3. Why might these trends be a problem?

____

____

____

Name: ______________________ Date: ______________

Directions: Follow the steps to complete the map. Then, answer the question.

1. Circle the capital of France.
2. Draw a + where the coasts are being added to.
3. Draw a – where the coasts are being worn a way.
4. Draw a box around cities that may be affected by these trends.
5. Should the people who live in cities away from the coast be concerned about coastal erosion and aggradation? Why or why not?

____

____

Read About It

Name: ______________________________ **Date:** ________________

Directions: Read the text, and study the photo. Then, answer the questions.

Coastal Erosion in France

Many French beaches are in danger. They are slowly eroding, or wearing away, into the ocean. This is harmful to beach ecosystems. Erosion can also be a problem for people. As the beach wears away, the ocean creeps closer to homes and businesses.

One popular way to reduce erosion is to build seawalls. These are long structures that jut into the water near a beach. Seawalls can stop big waves from crashing on the shore and eroding the beach. But the walls are very expensive to build and maintain. They can disrupt sea life. And in some cases, the walls stop working over time.

This seawall in France helps stop coastal erosion.

In France, choosing to build seawalls is up to individual regions. The central government does not help. This means the regions do not have to address the problem, and if they do, they have to pay for it themselves. So, the different regions have made different choices. Their beaches reflect their choices.

1. Why are French beaches in danger?

____

____

____

2. What are the advantages and disadvantages of building seawalls?

____

____

____

3. If you were in charge of deciding whether to build a seawall, what would you do? Why?

____

____

Name: ______________________________ Date: ______________

Directions: These photos show three different types of beaches in France. Study the photos, and answer the questions.

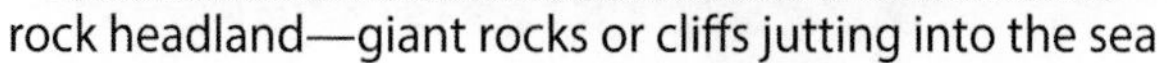
rock headland—giant rocks or cliffs jutting into the sea

open beaches—sandy beaches

mudflats and saltmarshes—marshy lands which are below the sea during high tide

1. Which beach do you think erodes most easily? Why?

____

____

____

2. How are the three beaches different?

____

____

3. Which type of beach do you think people are most eager to protect? Why?

____

____

Name: ______________________________ **Date:** ______________

Directions: Coasts in the United States are also eroding. Do you think all the states should help pay for seawalls or just the states affected by erosion? Write a persuasive paragraph stating your opinion. Include at least three supporting reasons.

Name: ______________________________ **Date:** ________________

Directions: The map shows three small countries in Europe. They are sometimes called the Baltic States. Study the map, and use the distance scale to answer the questions.

1. About how many kilometers is it from Tallinn, Estonia, to Vilnius, Lithuania?

____

2. About how many kilometers wide is Latvia at its widest point?

____

3. About how many kilometers is Riga, Latvia, from the Russian border?

____

Name: ____________________ Date: ____________

Creating Maps

Directions: Use the clues and the distance scale to label the missing cities on the map.

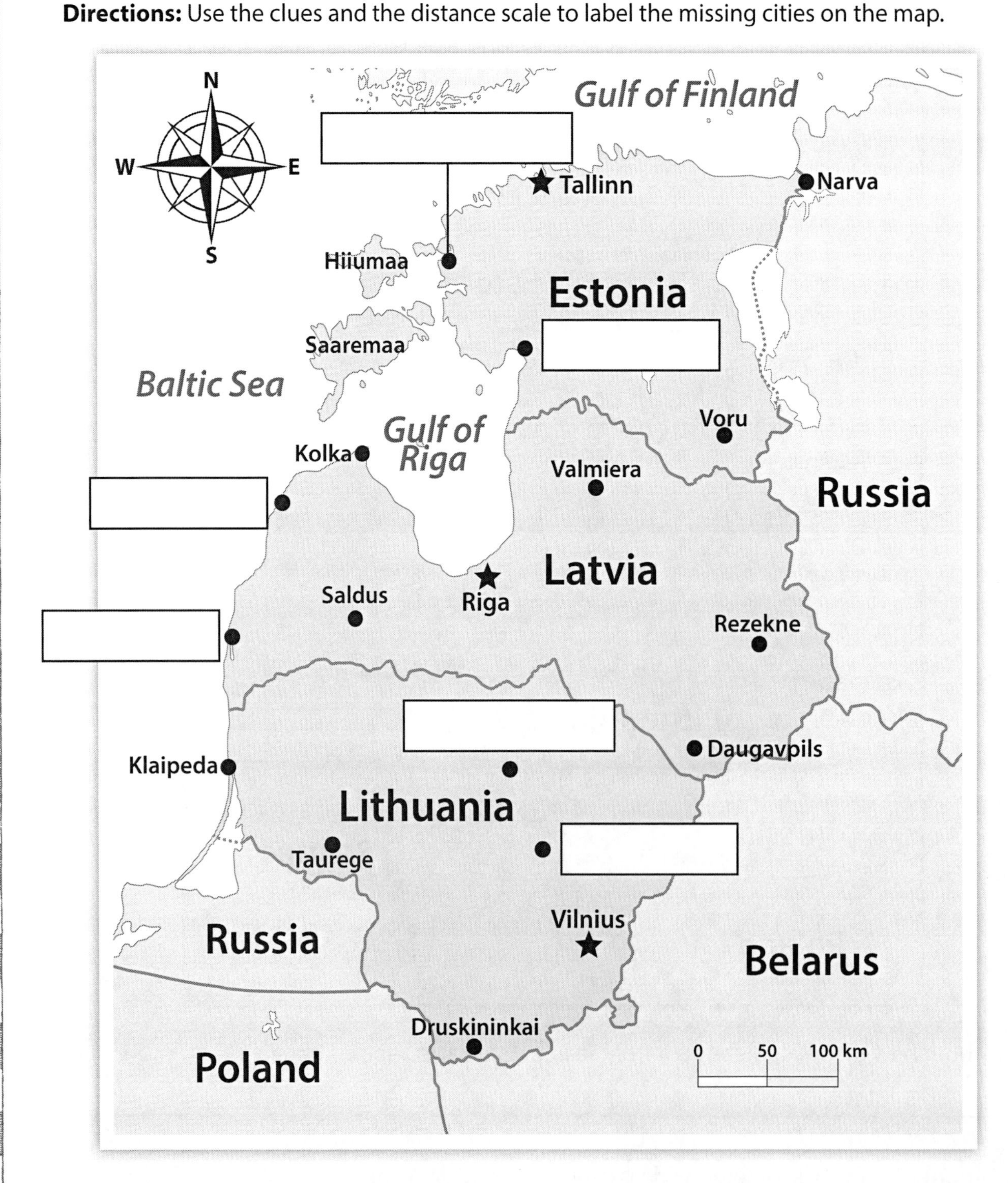

1. Haapsalu is 100 km southwest of Tallinn.
2. Panevezys is 150 km northwest of Vilnius.
3. Liepaja is 200 km southwest of Riga.
4. Ventspilis is 150 km northwest of Riga.
5. Ukmerge is 75 km northwest of Vinius.
6. Pärnu is 100 km south of Tallinn.

Name: ______________________ Date: ______________

Directions: Read the text, and study the photo. Then, answer the questions.

The Baltic States

Three small countries make up the Baltic States. They got their name because they border the Baltic Sea. They are in northeastern Europe. Of the three, Estonia is the farthest north. Latvia is in the middle. Lithuania is the farthest south. Latvia and Lithuania are almost the same size. Their size is equal to West Virginia in the United States. Estonia is even smaller.

The Baltic States have a similar recent history. All three were made part of the U.S.S.R. (Russia) in 1940. They remained under its control until 1991. Now, they each govern themselves. They are also a part of the European Union. Some people in the Baltic States speak Russian, but most speak their native languages. Each country was able to keep its own language and culture during Russian rule.

1. What are the names of the Baltic States?

____

2. How are their histories similar?

____

____

____

____

3. Why might some people in all three countries speak Russian?

____

____

____

____

Think About It

Name: ______________________ **Date:** ______________

Directions: These graphs show the first languages of people in the Baltic States. Study the charts, and answer the questions.

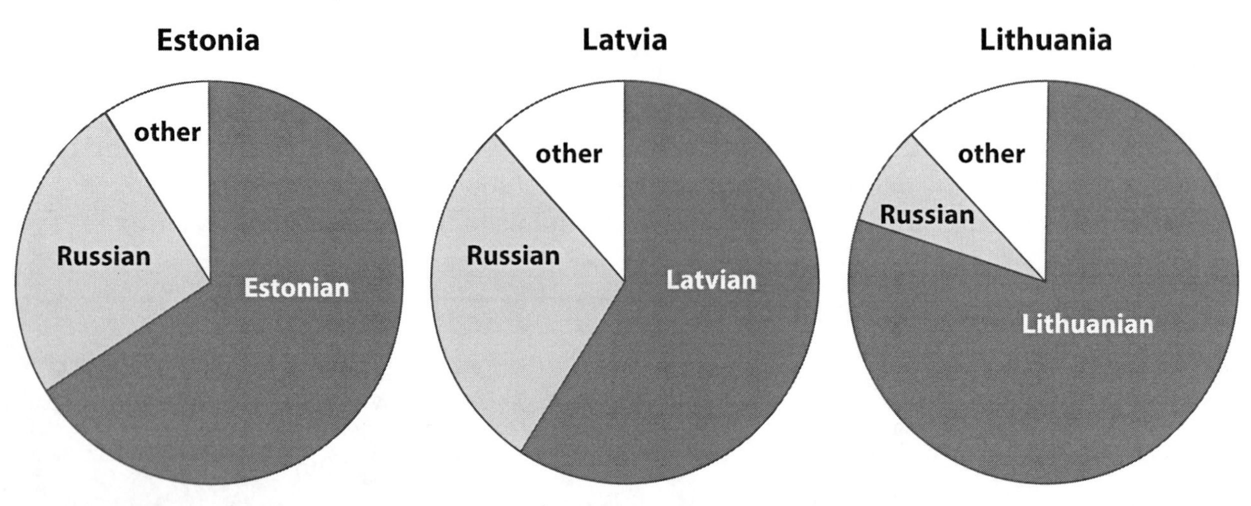

1. What languages do most people speak in each of the three countries?

____

____

2. What do these graphs tell you about the people of this region?

____

____

____

3. Many people from the Baltic States also speak English. How might that be helpful?

____

____

____

4. People in the Baltic States have held onto their own cultures. How do these graphs support that fact?

____

____

____

Name: ______________________________ Date: ______________

Directions: List six languages spoken by people in your country. Then, draw a picture of your community.

______________________ ______________________

______________________ ______________________

______________________ ______________________

Reading Maps

Name: ______________________ Date: ______________

Directions: Study the map of the Scandinavian countries. Then, answer the questions.

Scandinavian Countries

1. Describe the relative locations of these three countries in Europe.

___

___

___

2. Copenhagen is the capital of which country?

___

3. Name all countries or bodies of water that make Sweden's borders.

___

___

Name: ______________________________ **Date:** ______________

Directions: Read the text in the box, and study the legend. Add the correct symbol to each Scandinavian country. Then, shade each Scandinavian country a different color.

Scandinavian Resources

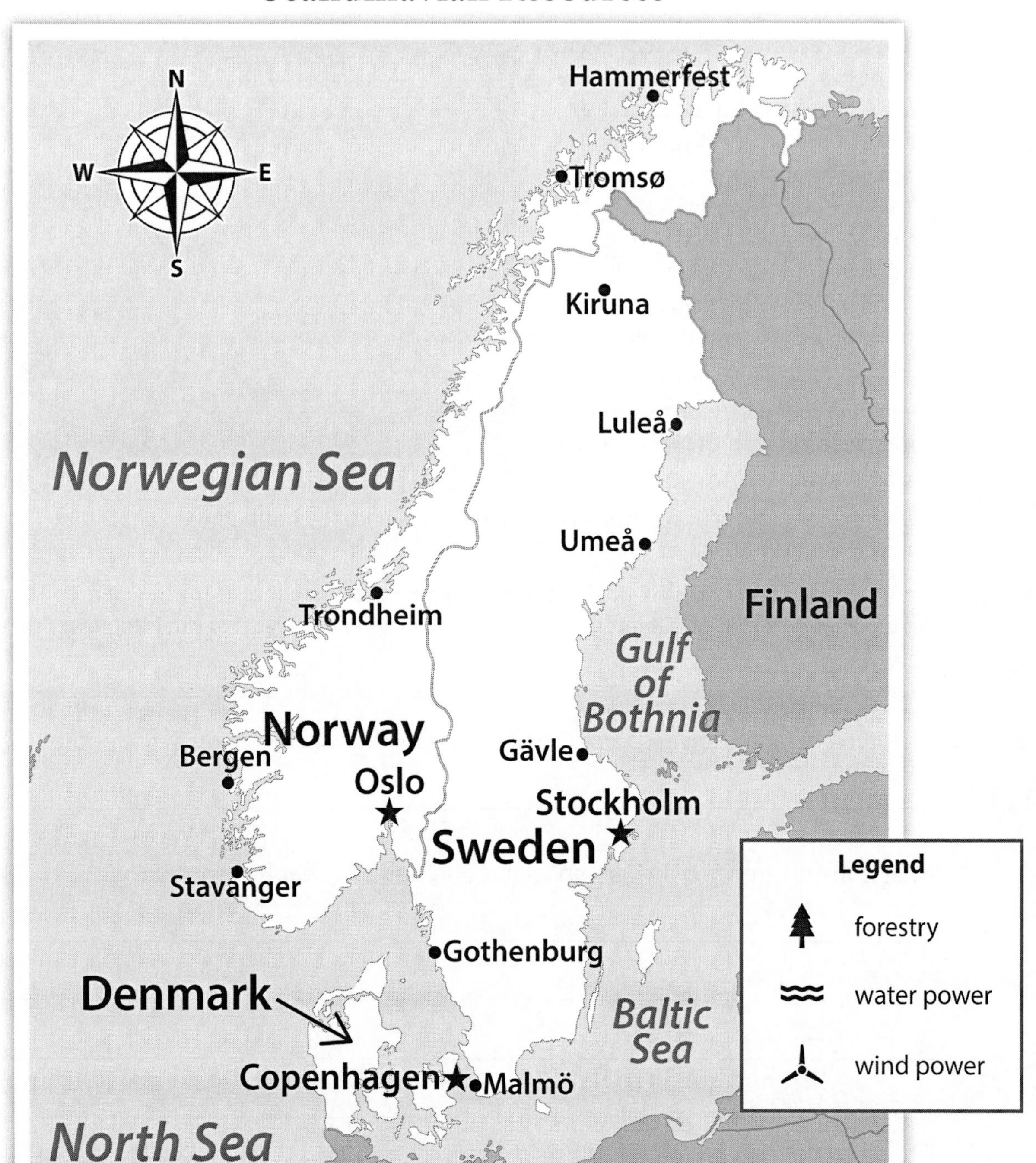

Scandinavian countries use many different resources to create power. But they are all trying to use more renewable resources. Denmark uses its high winds. Norway uses water to create power. Sweden uses its vast forests.

Read About It

Name: ______________________ **Date:** ______________

Directions: Read the text, and study the photo. Then, answer the questions.

Renewable Resources

We use resources every day. Some of the resources are renewable. These do not run out and can be used over and over. Others are non-renewable. This means that once they are used up, they are gone forever. Both types can be used to make electricity. But Scandinavian countries are trying to use more renewable resources. This reduces their impact on the planet.

Denmark has very fast winds. So people built many wind turbines. These structures turn the wind's energy into electricity.

Norway has access to waterpower. People in this country build power plants that harness the power of water.

Over half of Sweden is covered in forests. People use these forests for bioenergy. They take the energy stored in plants and turn it into electricity. People can use it to power their homes.

1. What type of renewable resource is being used by the wind turbines in the photo?

____

2. What is the difference between renewable and non-renewable resources?

____

____

____

____

3. What is bioenergy?

____

____

____

Name: ______________________________ **Date:** ________________

Directions: This graph shows how much wind energy Denmark plans to use for electricity. Study the graph, and answer the questions.

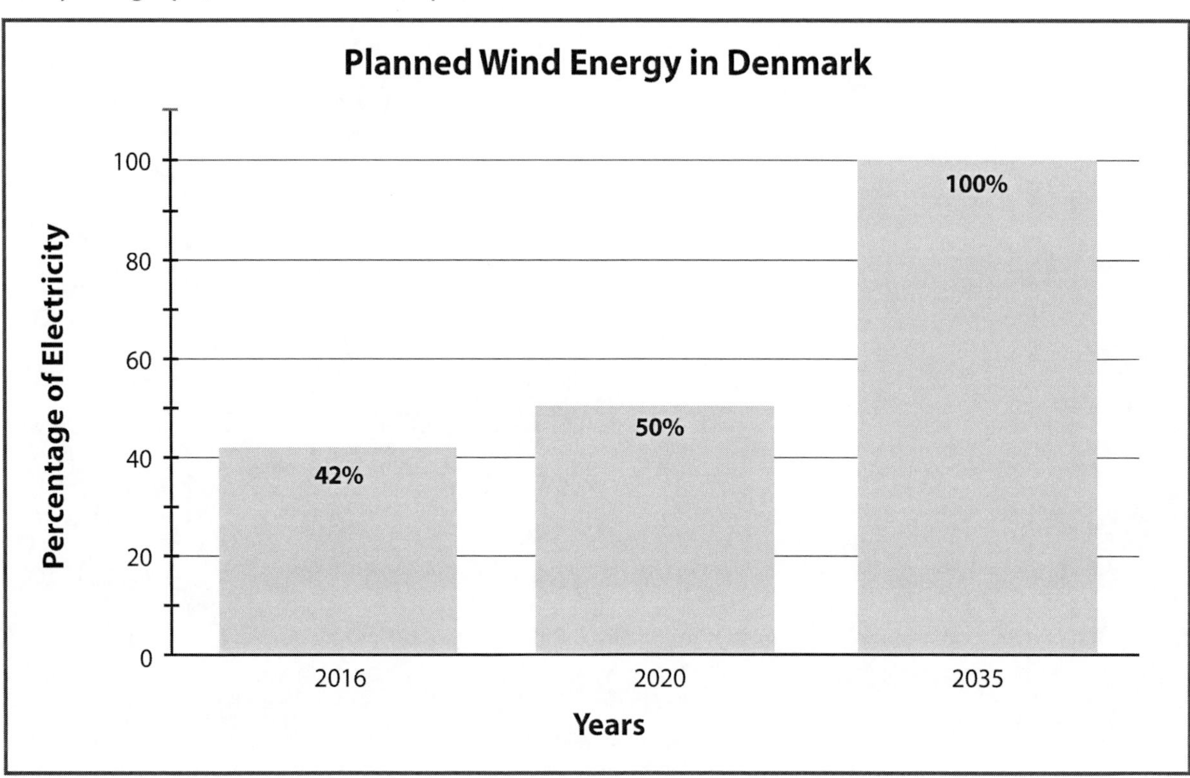

1. Why would Denmark want to use 100 percent renewable energy?

____

____

____

2. Do you think other countries use 100 percent renewable energy, too? Why or why not?

____

____

____

3. What could Denmark's next goal be after it reaches 100 percent renewable energy?

____

____

____

Name: ______________________ Date: ____________

Directions: How could your community save energy? How could it use more renewable resources? Draw and write to create a plan for your community.

Name: ______________________________ **Date:** ________________

Directions: The United Kingdom is made up of regions called countries. But, they are all part of the nation called the United Kingdom. Study the map, and answer the questions.

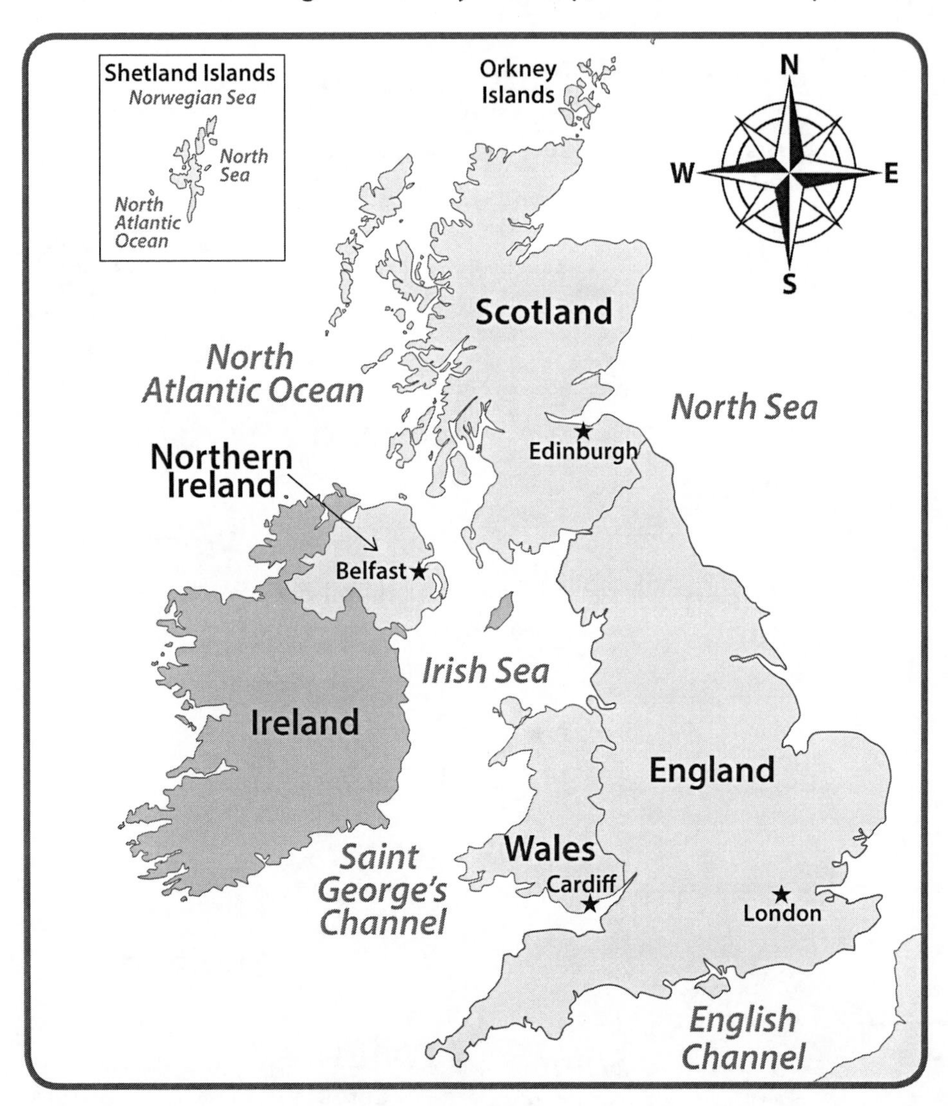

1. What four countries make up the United Kingdom?

____

____

____

2. What body of water is south of England?

____

3. Which country in the United Kingdom is the farthest north?

____

Name: ______________________________ Date: ________________

Directions: The information below describes where some of the natural resources the United Kingdom exports are found. Use the legend and the information to add the symbols the map.

British Natural Resources

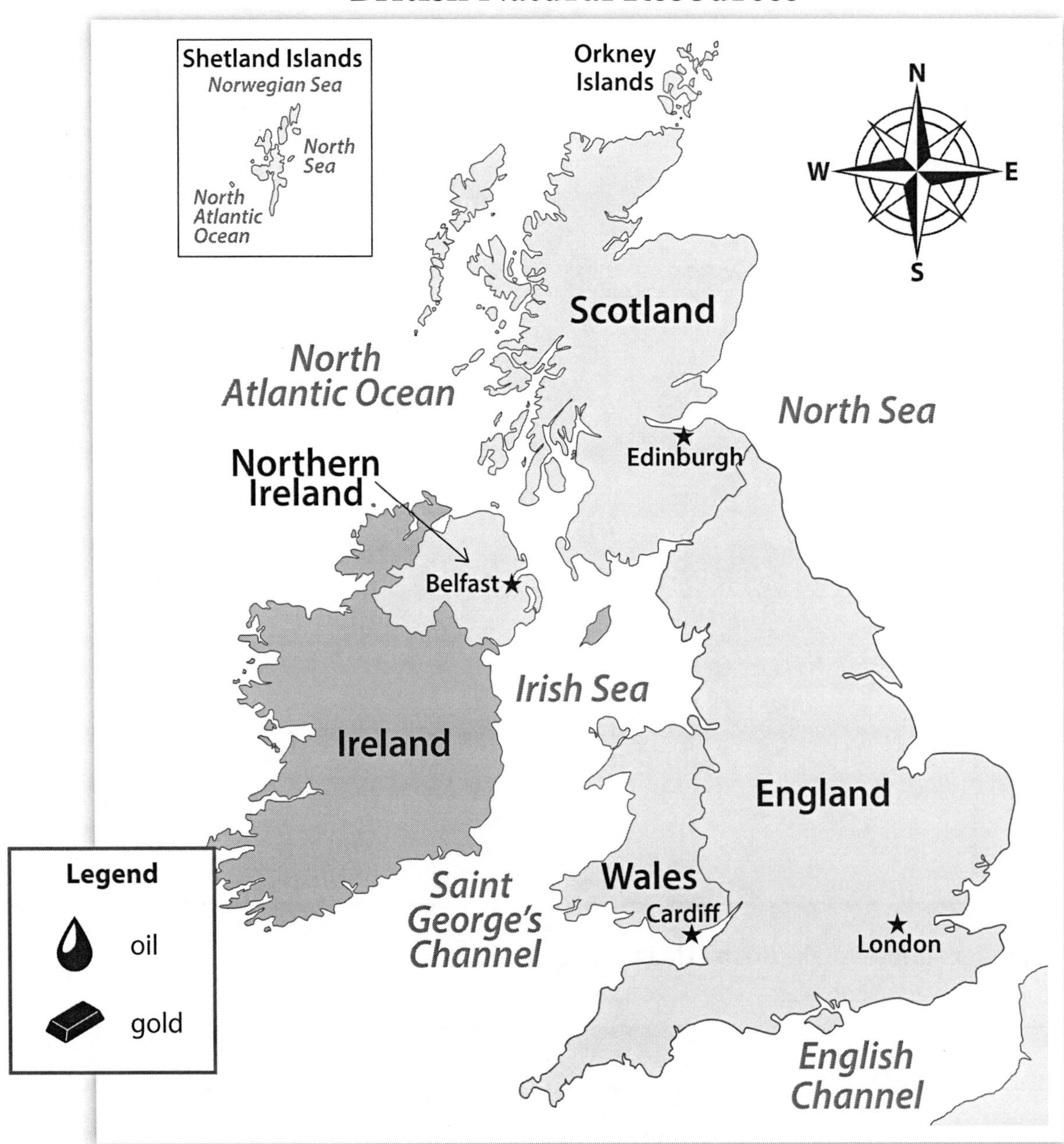

- The Hurricane Oil Field is in the Norwegian Sea just west of the Shetland Islands.
- The Forties Oil Field is in the North Sea, off the east coast of Scotland.
- Gold has been found in many places, including the southwest tip of England, the southern part of Northern Ireland, and just west of London, England.

Name: ______________________ Date: ______________

Directions: Read the text, and study the photo. Then, answer the questions.

Trade During the Industrial Revolution

A great change started in the United Kingdom in the late 1700s. It was called the Industrial Revolution. New machines and tools were invented. Weaving fabric from cotton and wool became faster and easier. People found a better way to cast iron. This made it easier to make tools and weapons. The invention of the steam engine changed the way people travel.

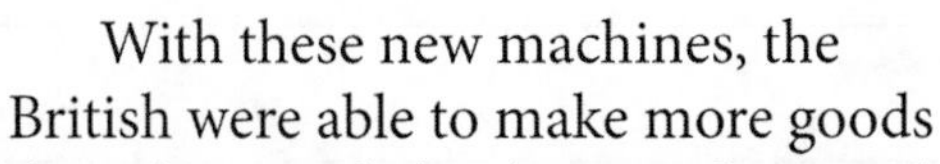

With these new machines, the British were able to make more goods than they needed. The British started trading more with other nations. They traveled to India to trade for tea and spices. They traded with the United States for cotton. The British traded with the Caribbean Islands for sugar. The United Kingdom's economy grew because of their imports and exports.

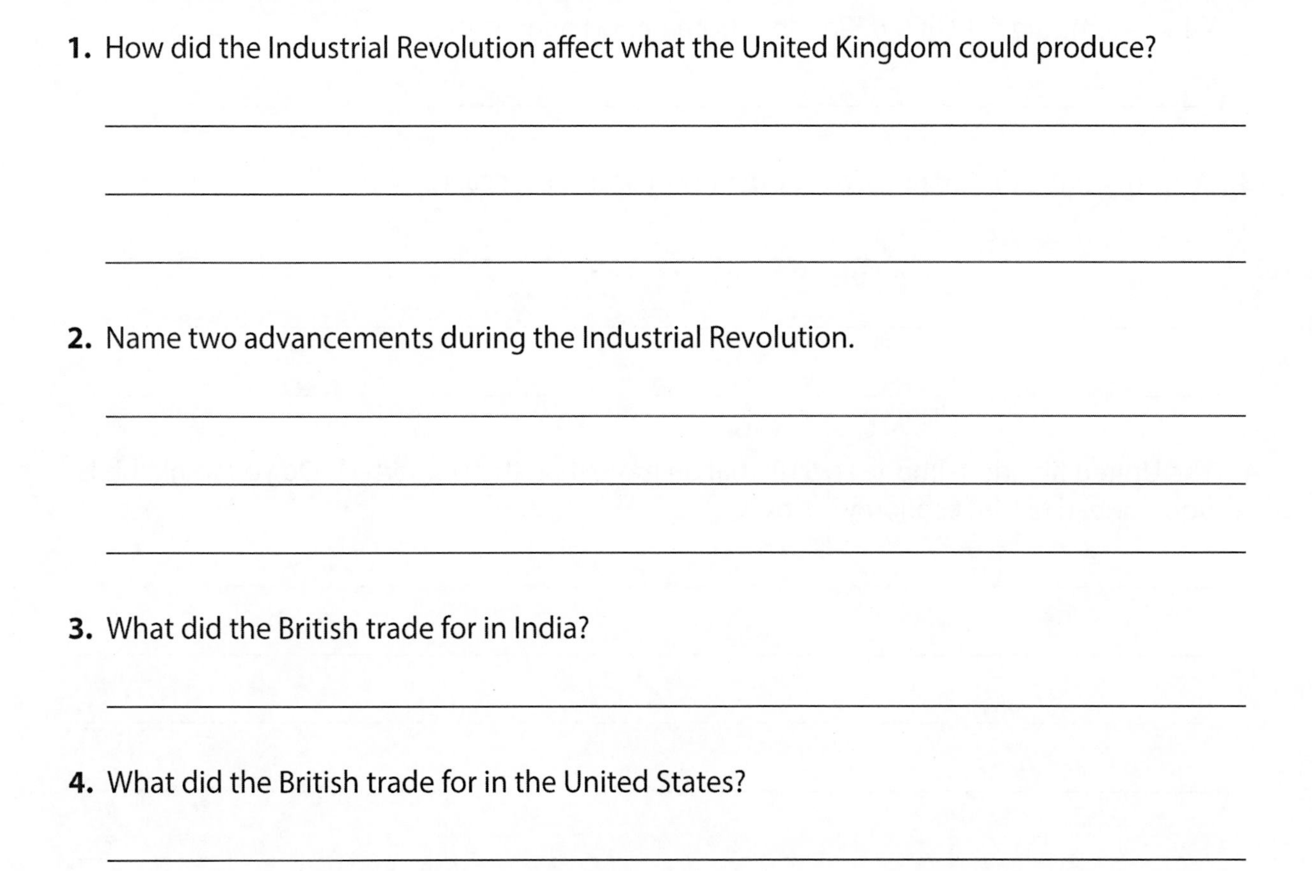

1. How did the Industrial Revolution affect what the United Kingdom could produce?

____

____

____

2. Name two advancements during the Industrial Revolution.

____

____

____

3. What did the British trade for in India?

____

4. What did the British trade for in the United States?

____

Think About It

Name: ______________________ **Date:** ______________

Directions: This chart shows some of the United Kingdom's top imports and exports in 2015. Imports are things a country buys from other countries. Exports are things a country sells to other countries. Study the chart, and answer the questions.

Imports	Exports
gold	cars
cars	medicine
medicine	oil
gas turbines	aircraft parts
oil	vehicle parts

1. Why might the United Kingdom need to import items?

____

____

____

____

2. Which items did the United Kingdom both import and export?

____

3. Why might the United Kingdom both import and export cars?

____

____

____

4. The United Kingdom imported $181 billion more than they exported. Do you think this is good or bad for the economy? Why?

____

____

____

____

____

Name: ______________________________ Date: ________________

Directions: New inventions give countries new things to import and export. Use facts and your imagination to draw and describe inventions of the past, present, and future.

Inventions of the Past

Modern Inventions

Inventions of the Future

Name: ______________________ Date: ______________

Directions: Study the map of Greece. Then, answer the questions.

1. Name four island cities that are part of Greece.

____

____

____

2. Kavala is on the coast of which sea?

____

3. What is the capital of Greece?

____

4. What landforms do you see in Greece?

____

Name: ______________________________ **Date:** ______________

Directions: Use the clues to label the countries that border Greece. Then, choose a color and shade Greece. Outline the countries Greece borders in different colors. Make a legend to show what your color represents.

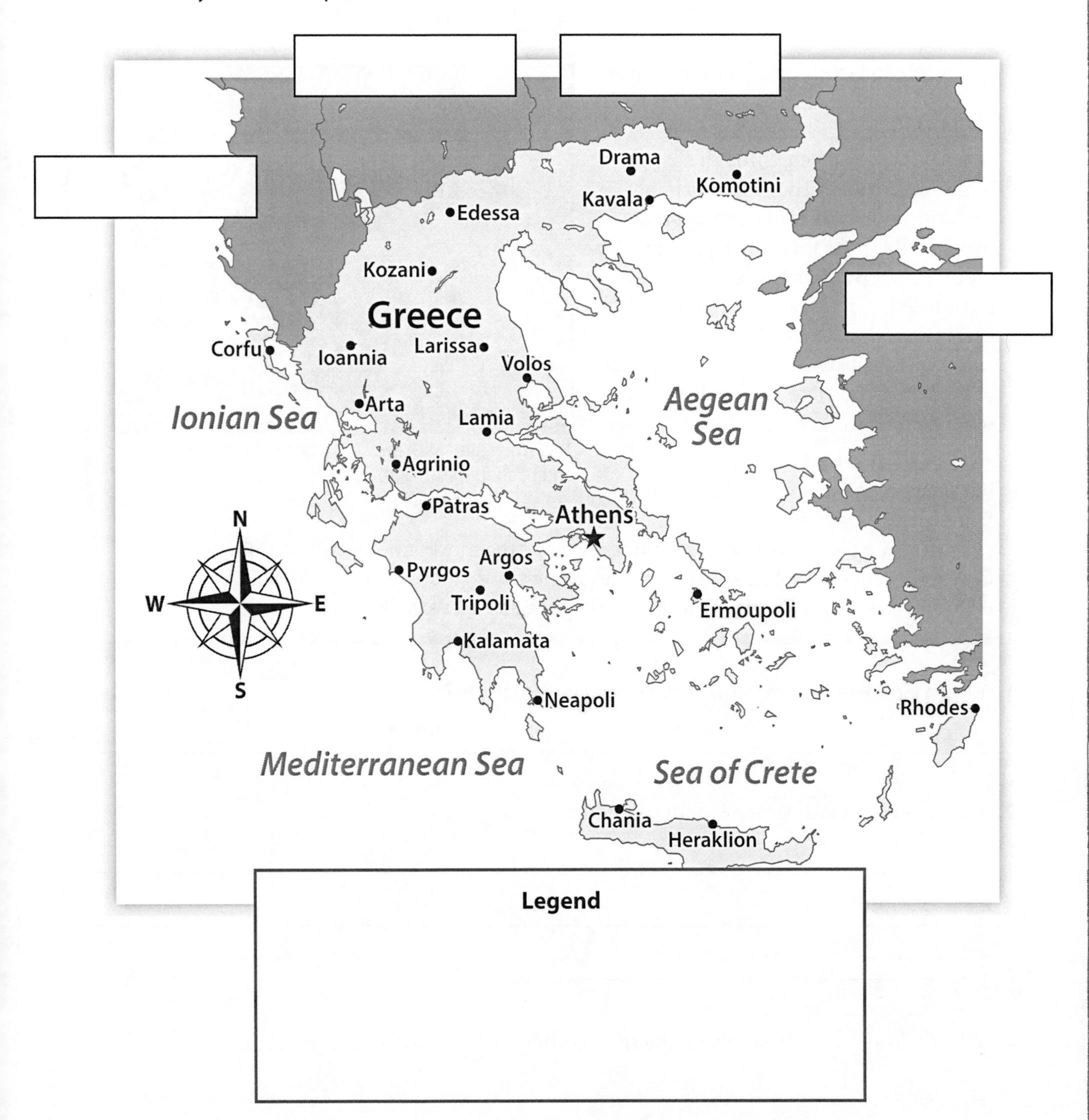

Legend

- Bulgaria is north of Komotini.
- Macedonia is north of Edessa.
- Turkey is east of Ermoupoli.
- Albania is northwest of Ioannina.

Name: ______________________________ **Date:** ____________________

Directions: Read the text, and study the photo. Then, answer the questions.

The Greek Alphabet

People in Greece speak Greek, which is one of the oldest languages in the world. It has been spoken for over 3,400 years! Greek has gone through changes in its long history. The oldest form of the language is ancient Greek. At first, people only spoke this language. Then, around 1000 BC, the Greek alphabet was introduced. Greek has not changed much since the 16th century. This form of the language is called Modern Greek. Some of the letters look like English letters, but others are very different.

This street sign in Greece is written in both the Greek and Latin (English) alphabet.

Most Americans cannot speak Greek, but many of them have used its alphabet. Some Greek letters are used in mathematics to represent an idea or unknown number. One of the most famous is the number pi. In Greek, the letter pi looks like "Π". In college, there are groups people can join. A fraternity is for men, and a sorority is for women. They are named with Greek letters.

1. Which Greek letters are similar to English in the photo?

____

2. How are Greek letters used in mathematics?

____

____

____

3. Why do you think the Greek language has changed so much?

____

____

____

Name: ______________________ **Date:** ______________

Directions: This chart shows how Greek words are part of some English words. Study the chart, and answer the questions.

Greek Word	Meaning	English Word	Meaning
autos	self	automatic	working by it**self** without help
bios	life	biology	the study of plant and animal **life**
grapho	write	paragraph	a small section of **writing** about one idea
treis (*tri*)	three	triangle	a shape with **three** sides
anti	opposite, against	antifreeze	liquid that keeps water in engines from freezing

1. How can knowing these Greek words help you learn new English words?

____

____

____

____

2. Use the chart to explain what the word *autobiography* means.

____

____

____

3. What is another English word that uses the Greek word *tri*?

____

4. What is another English word that uses the Greek word *anti*?

____

Name: ______________________________ Date: ____________________

Directions: Study the chart showing the Greek alphabet. Then, write letters that are found in both the Greek and English alphabets in the middle. Write letters that are only in Greek on the right. Write the rest of the English alphabet on the left.

Α = alpha	Ι = iota	Ρ = rho
Β = beta	Κ = kappa	Σ = sigma
Γ = gamma	Λ = lambda	Τ = tau
Δ = delta	Μ = mu	Υ = upsilon
Ε = epsilon	Ν = nu	Φ = phi
Ζ = zeta	Ξ = xi	Χ = chi
Η = eta	Ο = omicron	Ψ = psi
Θ = theta	Π = pi	Ω = omega

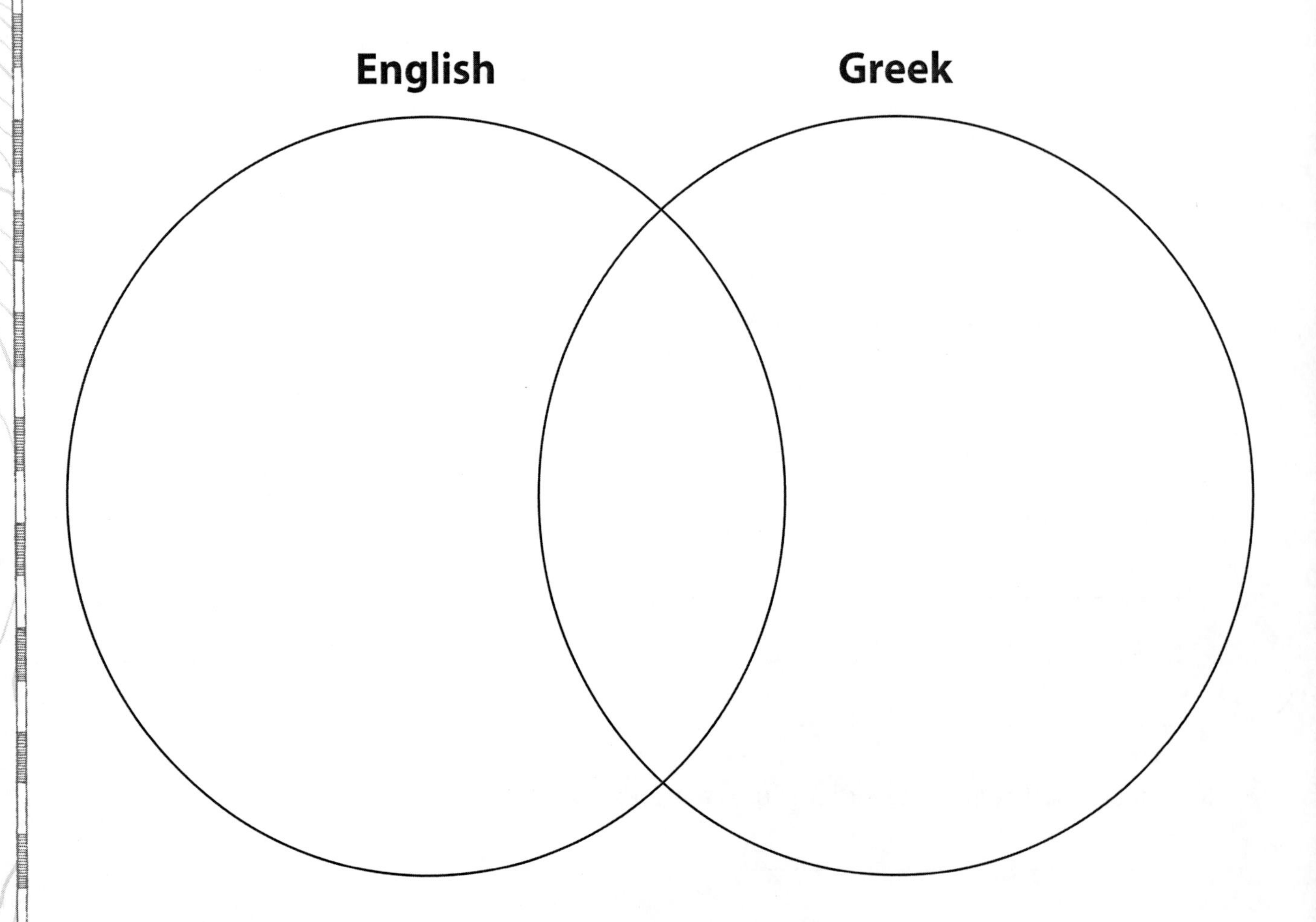

Name: ______________________________ Date: ____________________

Directions: Study the map of the world. Then, answer the questions.

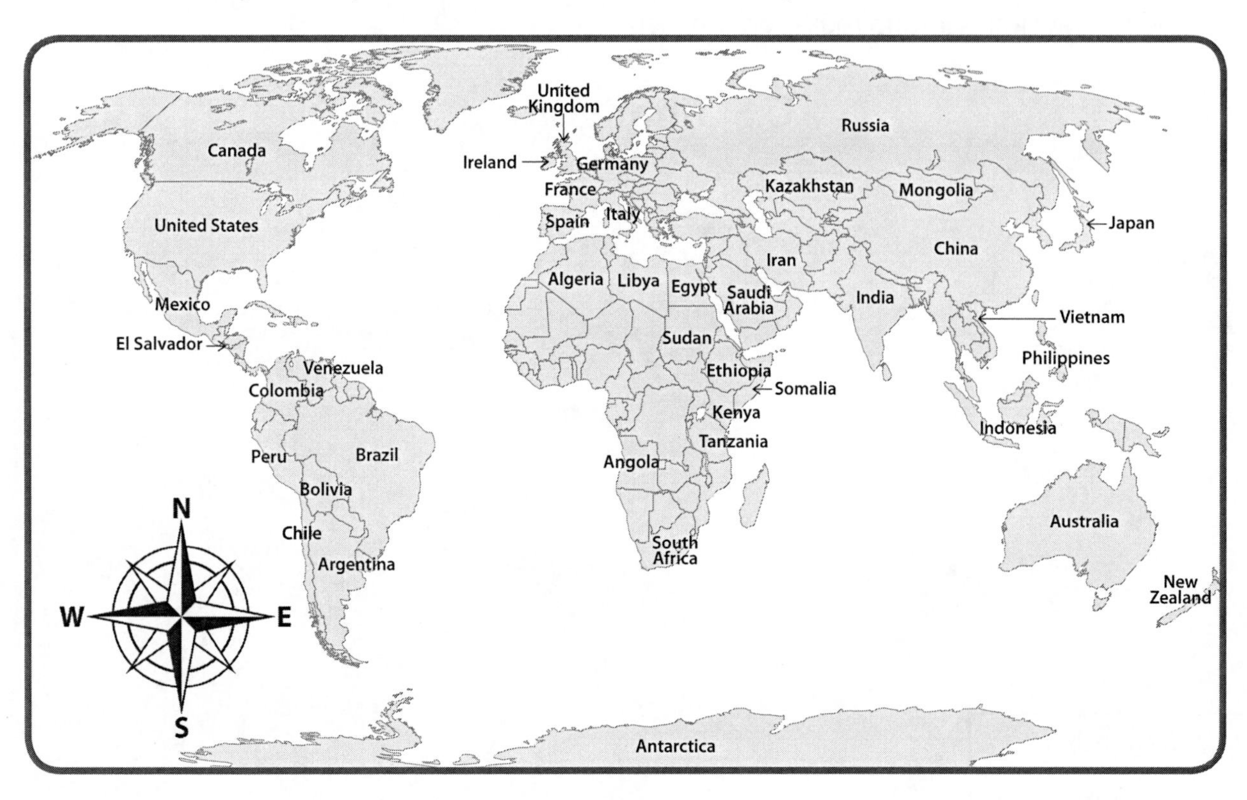

1. What country is south of Russia and north of China?

____

2. Name one country that borders Russia.

____

3. Name three countries in North America.

____

____

____

4. Name three countries in Asia.

____

____

____

Name: ______________________________ Date: ________________

Directions: The text below describes immigration to the United States. Use the text and the legend to draw immigration routes on the map.

U.S. Immigration

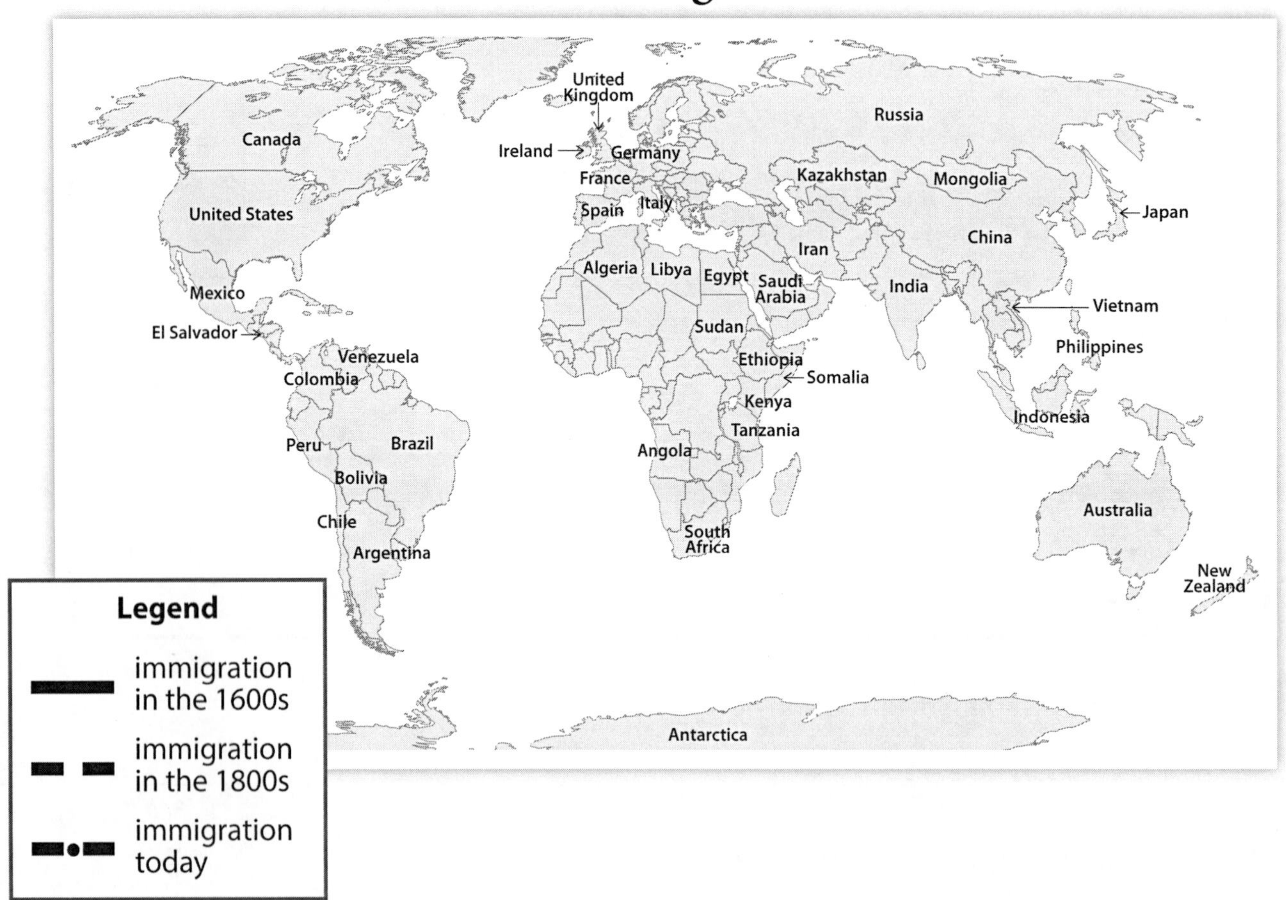

- The first groups of immigrants came from England (the United Kingdom) in the 1600s.
- In the 1800s, most immigrants came from western European countries, such as the United Kingdom, Ireland, and Germany. But small numbers of immigrants began coming from Japan and China.
- Today, many immigrants come to the United States from Mexico, China, India, the Philippines, Vietnam, and El Salvador.

Challenge: Draw one or more routes on the map to show where your family immigrated from.

Name: ______________________ Date: ______________

Directions: Read the text, and study the photo. Then, answer the questions.

Reasons to Immigrate

When people move from one country to another, they are called *immigrants*. The United States is a country made of immigrants. This makes it a unique blend of cultures. But why would someone leave home and move to a different country? There are several reasons.

Sometimes, people cannot practice their religion in their home countries. This brought the Pilgrims to the New World in 1620. They moved so they could have religious freedom. Others come because of a famine or lack of work in their home countries. In the 1840s, there was a potato famine in Ireland. Millions of people moved to the United States because there was not enough food for them in Ireland. Some people come because of war. They might not be safe in their home countries. Today, many people come to the United States from Syria and Somalia because of war. Others just want to pursue new opportunities. People come to the United States from all over the world to build better lives for themselves.

1. What is an immigrant?

____

____

2. Why did people from Ireland move to the United States in the 1800s?

____

____

____

3. Why would war make people want to immigrate?

____

____

Think About It

Name: ______________________ **Date:** ______________

Directions: This chart shows the number of immigrants that came to the United States in different years. Study the chart, and answer the questions.

Year	Number of Immigrants
1830	23,322
1860	153,640
1890	455,302
1920	430,001
1950	249,187
1980	524,295
2010	1,042,625

1. Which two years had the biggest jump in the number of immigrants?

____

2. What do you notice about the number of immigrants in 1950? How is this different from other years? Why do you think this happened?

____

____

____

____

____

3. What do you think the number of immigrants might be in 2040? Why?

____

____

____

____

Challenge: Create a graph on a separate sheet of paper to show the data from the chart.

Name: ______________________________ **Date:** ____________________

Directions: Imagine you are moving to a new country. People in the new country speak a different language and have a very different culture than yours. Write a fictional narrative to describe your experience.

Reading Maps

Name: ______________________ **Date:** ______________

Directions: Study the map of the world, and answer the questions.

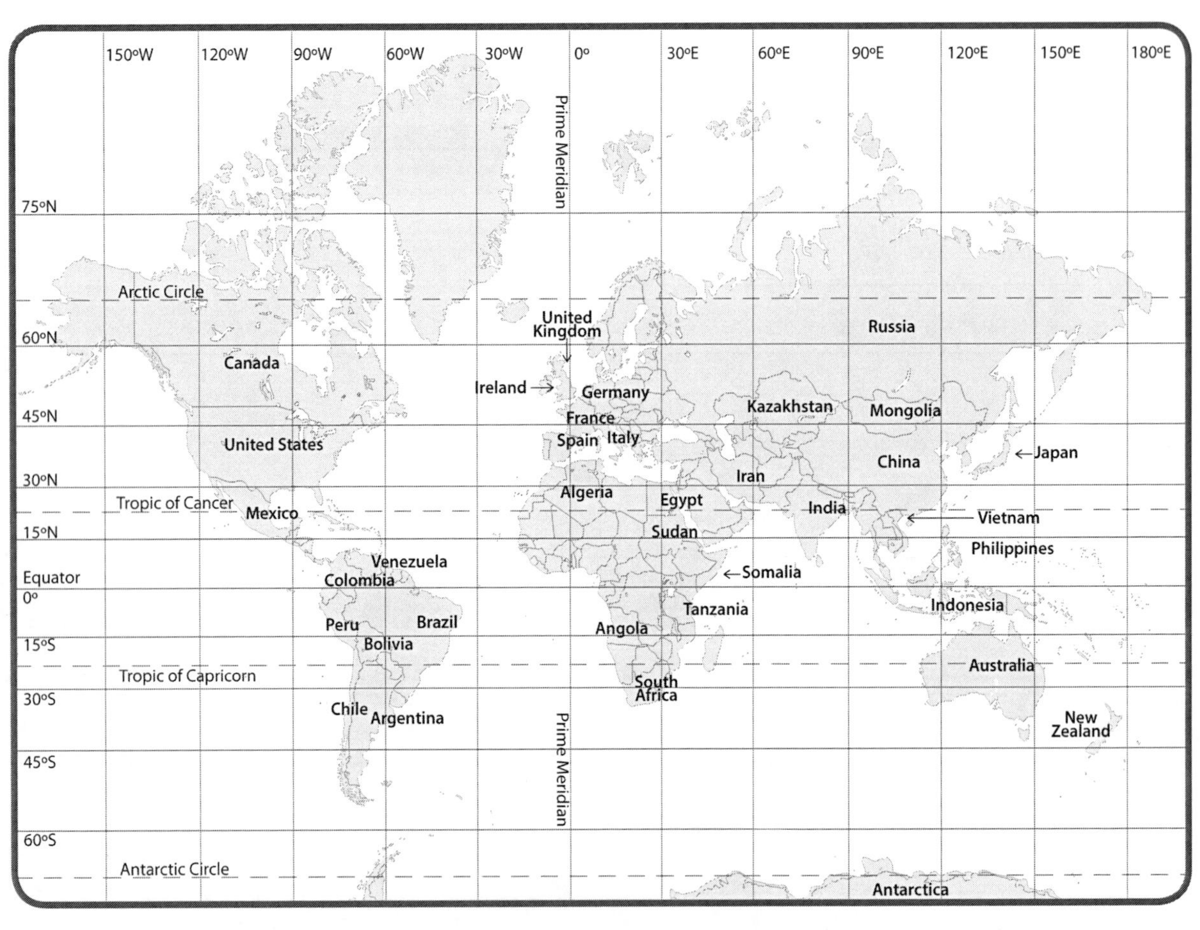

1. What country is located at 15°S, 45°W?

____

2. Give a coordinate in China.

____

3. Find the 120°E longitude line. What are two countries it passes through?

____

4. Find the 60°N latitude line. What are two countries it passes through?

____

5. Give a coordinate in Australia.

____

Name: ______________________________ **Date:** ____________

Directions: Follow the steps to complete the map.

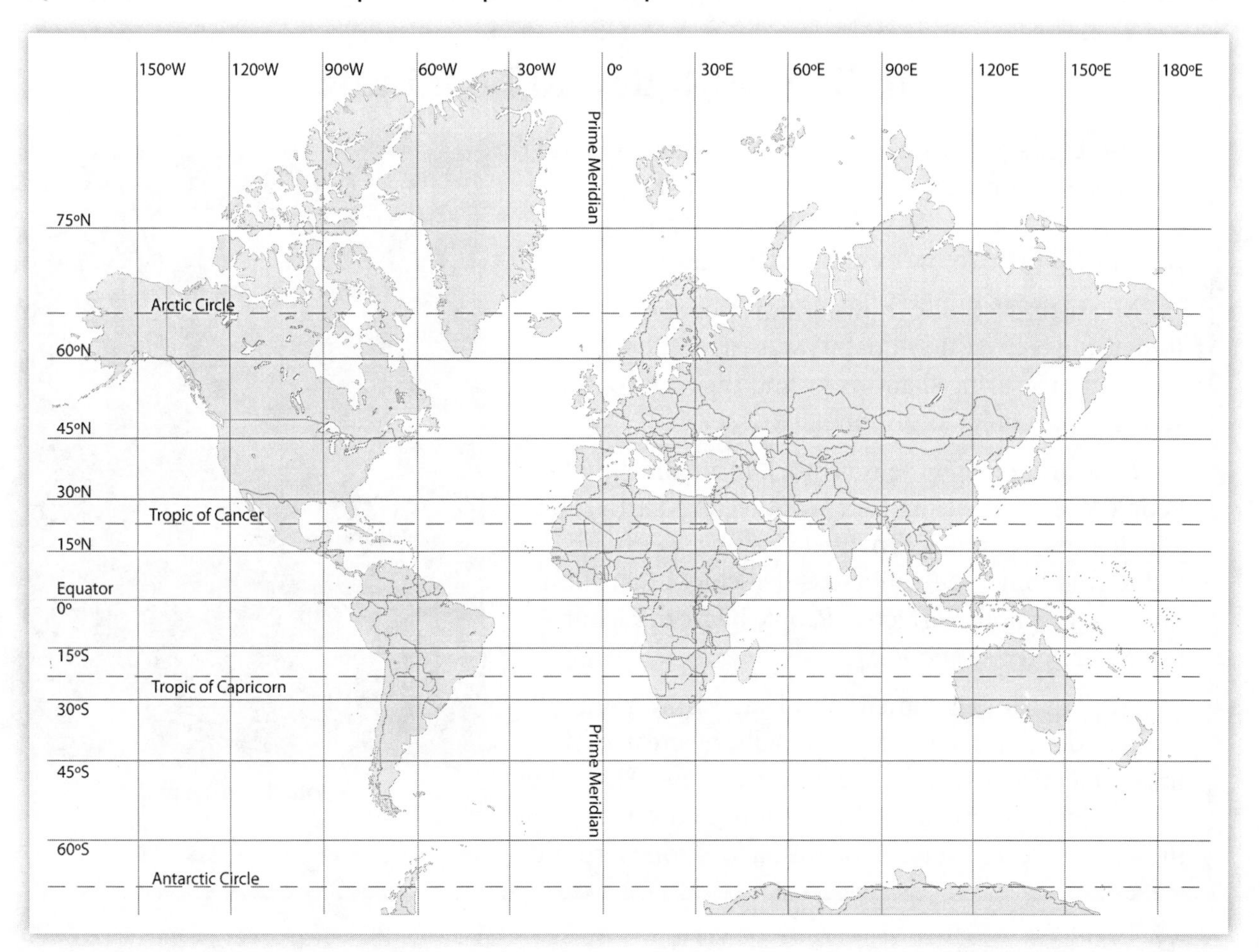

1. Write an *A* at 45°N, 120°W.
2. Write a *B* at 60°N, 105°E.
3. Write a *C* at 30°S, 150°W.
4. Write a *D* at 15°N, 75°E.
5. Write an *E* at 15°S, 45°E.
6. Trace the equator with a red crayon.
7. Trace the prime meridian with a green crayon.
8. Trace the Tropic of Cancer with a yellow crayon.
9. Trace the Tropic of Capricorn with an orange crayon.

Read About It

Name: ______________________ **Date:** ______________

Directions: Read the text, and study the photo. Then, answer the questions.

Finding Latitude and Longitude

Hundreds of years ago, there was no exact way to find absolute location. This put sailors in grave danger of getting lost at sea. Dividing the world into a grid would help them know exactly where they were.

Measuring latitude (how far north or south something is from the equator) was easy. It only required measuring shadows to find the angle of the sun. People could do this on land or sea.

Measuring longitude was much more difficult. People knew measuring time was the key. Earth rotates 360 degrees every day. That is 15 degrees per hour. But to find longitude, sailors needed clocks that could measure one hour perfectly. People had clocks, but none were precise enough.

An Englishman named John Harrison worked for years to develop such a clock. Finally, he created his masterpiece—a clock that kept perfect time. He chose Greenwich, England, as 0 degrees longitude. Longitude lines measure how far east or west a location is from there.

John Harrison

1. How did sailors find latitude?

____

____

2. Why was a clock that kept perfect time needed?

____

____

____

3. How do you think Harrison's clock changed navigation?

____

____

____

Name: ______________________________ **Date:** ________________

Directions: This map of North and South America was created in 1570. Study the map, and answer the questions.

1. Why are the longitude lines curved?

____

____

2. What differences do you notice between this map and modern maps?

____

____

____

3. What does this map tell you about exploration in the 1500s?

____

____

____

Name: ______________________________ **Date:** ______________

Directions: Use the map to answer the questions.

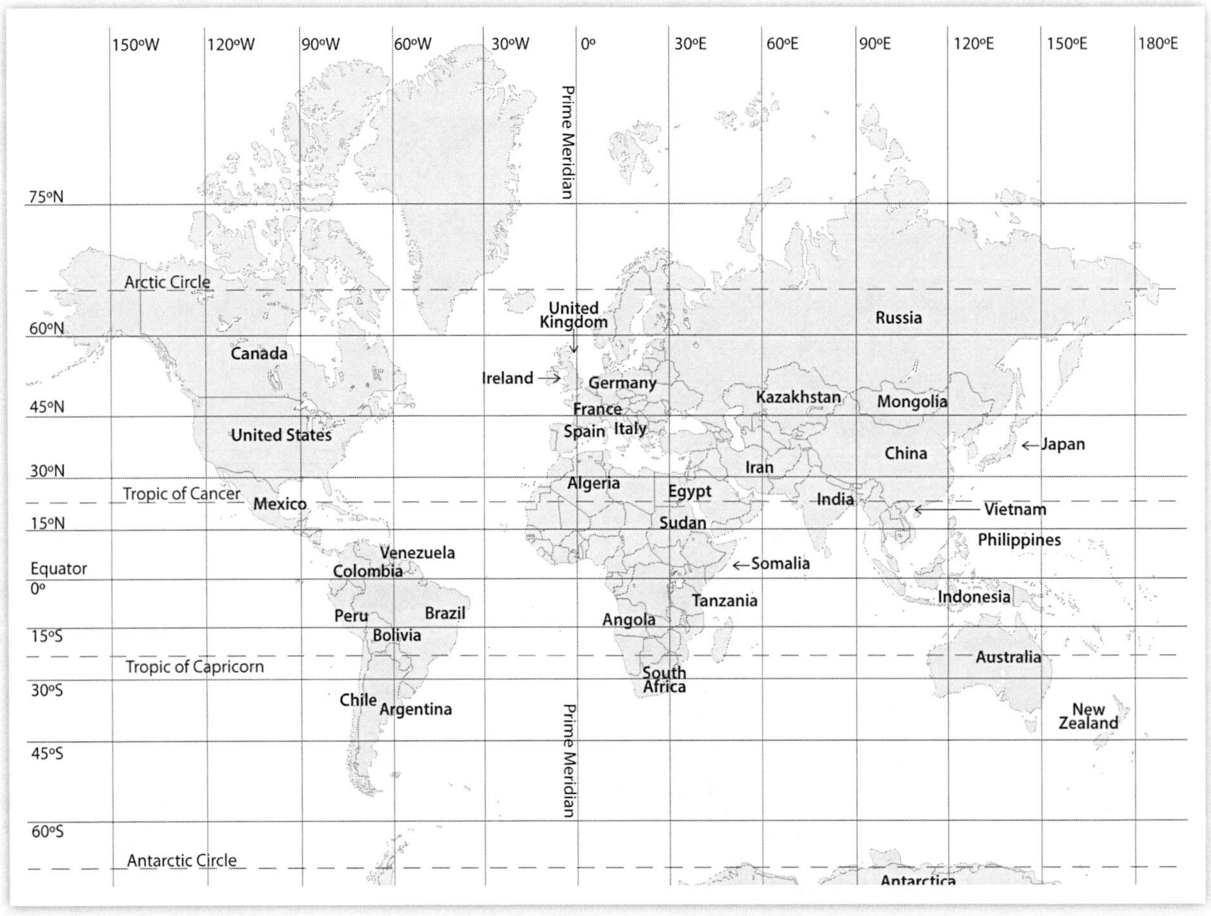

150°W
120°W
90°W
60°W
30°W
0°
30°E
60°E
90°E
120°E
150°E
180°E
Prime Meridian
75°N
Arctic Circle
60°N
45°N
30°N
Tropic of Cancer
15°N
Equator
0°
15°S
Tropic of Capricorn
30°S
45°S
60°S
Antarctic Circle
Canada
United States
Mexico
United Kingdom
Ireland
Germany
France
Spain
Italy
Russia
Kazakhstan
Mongolia
China
Japan
Iran
India
Vietnam
Philippines
Algeria
Egypt
Sudan
Somalia
Tanzania
Angola
South Africa
Indonesia
Australia
New Zealand
Venezuela
Colombia
Peru
Brazil
Bolivia
Chile
Argentina
Prime Meridian
Antarctica

1. What is the approximate latitude and longitude of where you live?

____

2. What is the latitude and longitude of a place in the United States you would like to visit?

place: ____

latitude/longitude: ____

3. What is the latitude and longitude of a place outside the United States you would like to visit?

place: ____

latitude/longitude: ____

Name: ______________________________ Date: ____________________

Directions: Study the map of the United States. Color the states you already know. Then, use a different color to shade the states you still need to learn.

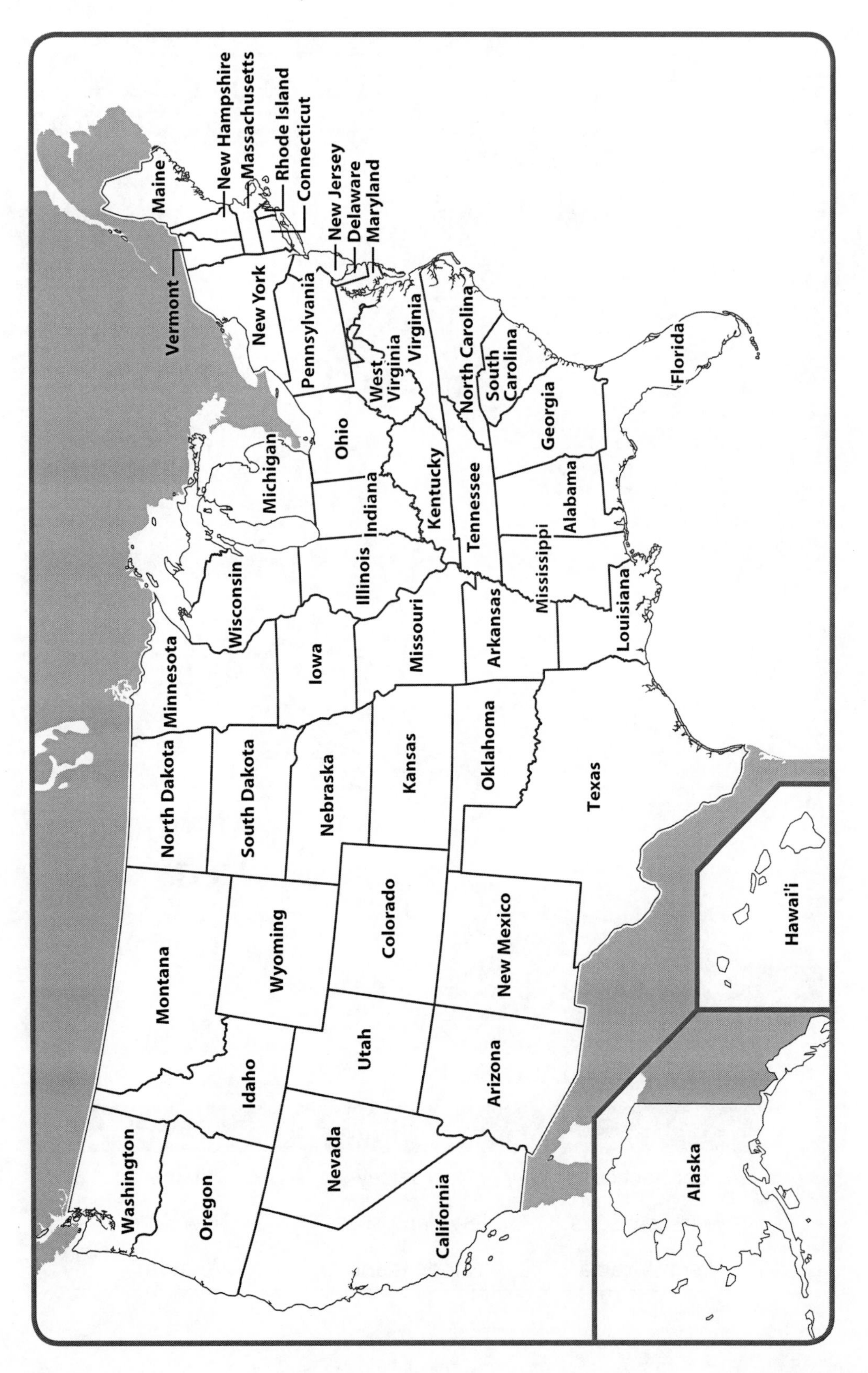

Name: ______________________________ **Date:** ________________

Directions: This is a map of the Northeast. Use the Word Bank to label the states.

Word Bank		
Connecticut	New Jersey	Maine
Massachusetts	New Hampshire	New York
Pennsylvania	Rhode Island	Vermont

Name: ______________________ **Date:** ______________

Directions: This is a map of the Midwest. Use the Word Bank to label the states.

Word Bank			
Illinois	Indiana	Iowa	Kansas
Michigan	Minnesota	Missouri	Nebraska
North Dakota	Ohio	South Dakota	Wisconsin

Name: ________________________________ **Date:** ____________________

Directions: This is a map of the South. Use the Word Bank to label the states.

Word Bank			
Alabama	Arkansas	Delaware	Florida
Georgia	Kentucky	Louisiana	Maryland
Mississippi	North Carolina	Oklahoma	South Carolina
Tennessee	Texas	Virginia	West Virginia

Name: ____________________ Date: ____________

Directions: This is a map of the West. Use the Word Bank to label the states.

Word Bank		
Alaska	Arizona	California
Colorado	Hawai'i	Idaho
Montana	New Mexico	Nevada
Oregon	Utah	Washington
Wyoming		

Name: ______________________________ Date: ______________

Directions: Study the state capitals. Color the states with capitals you already know. Then, use a different color to shade the states with capitals you still need to learn.

Name: ______________________ Date: ______________

Directions: Match each state with its capital.

1. Connecticut ______________
2. Iowa ______________
3. Kansas ______________
4. Maine ______________
5. Massachusetts ______________
6. Minnesota ______________
7. Missouri ______________
8. Nebraska ______________
9. New Hampshire ______________
10. New Jersey ______________
11. New York ______________
12. North Dakota ______________
13. Pennsylvania ______________
14. Rhode Island ______________
15. South Dakota ______________
16. Vermont ______________

Word Bank			
Albany	Augusta	Bismarck	Boston
Concord	Des Moines	Harrisburg	Hartford
Jefferson City	Lincoln	Montpelier	Pierre
Providence	Saint Paul	Topeka	Trenton

Name: ______________________ **Date:** ______________

Directions: Match each state with its capital.

1. Alabama ______________________

2. Arkansas ______________________

3. Florida ______________________

4. Georgia ______________________

5. Illinois ______________________

6. Indiana ______________________

7. Kentucky ______________________

8. Louisiana ______________________

9. Michigan ______________________

9. Mississippi ______________________

10. North Carolina ______________________

12. Ohio ______________________

13. Oklahoma ______________________

14. South Carolina ______________________

15. Tennessee ______________________

16. Texas ______________________

17. Wisconsin ______________________

Word Bank				
Atlanta	Austin	Baton Rouge	Columbia	Columbus
Frankfort	Indianapolis	Jackson	Lansing	Little Rock
Madison	Montgomery	Nashville	Oklahoma City	Raleigh
Springfield	Tallahassee			

Name: ______________________ Date: ______________

Directions: Match each state with its capital.

1. Alaska ______________________
2. Arizona ______________________
3. California ______________________
4. Colorado ______________________
5. Delaware ______________________
6. Hawai'i ______________________
7. Idaho ______________________
8. Maryland ______________________
9. Montana ______________________

9. Nevada ______________________
10. New Mexico ______________________
12. Oregon ______________________
13. Utah ______________________
14. Virginia ______________________
15. Washington ______________________
16. West Virginia ______________________
17. Wyoming ______________________

Word Bank				
Annapolis	Boise	Carson City	Charleston	Cheyenne
Denver	Dover	Helena	Honolulu	Juneau
Olympia	Phoenix	Richmond	Sacramento	Salem
Salt Lake City	Santa Fe			

Name: ______________________________ **Date:** ________________

Directions: Write the state capitals on the map.

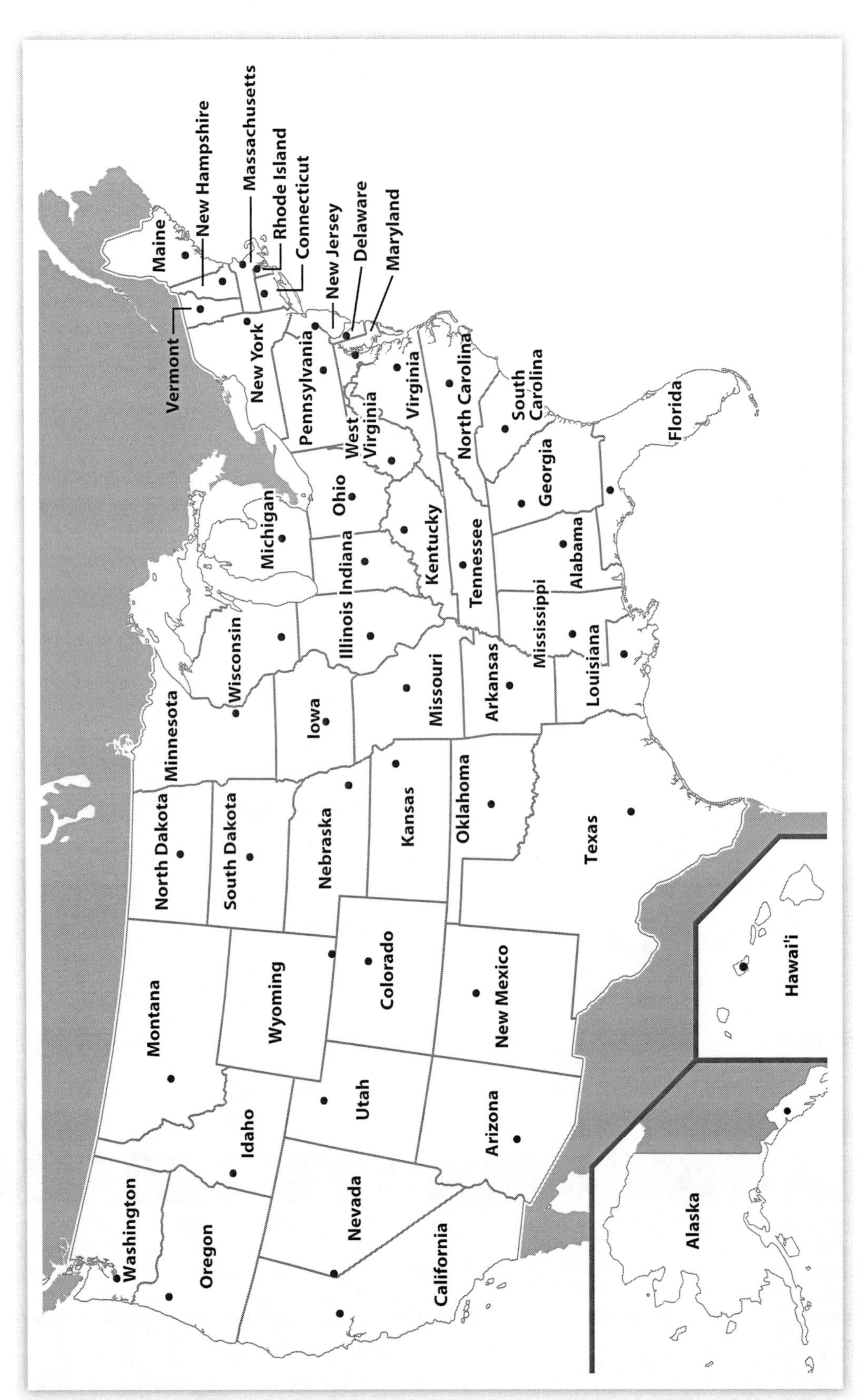

ANSWER KEY

There are many open-ended pages and writing prompts in this book. For those activities, the answers will vary. Answers are only given in this answer key if they are specific.

Week 1 Day 2 (page 16)

Students should use the compass rose to describe which continents and bodies of water are adjacent to their continent.
Continents on the equator: South America, Africa, and Asia; continents entirely south of the equator: Antarctica and Australia

Week 1 Day 3 (page 17)

1. A boat should be drawn in the river to the left of the bridge.
2. The path should cross the street and go over the bridge.
3. A fence should be drawn around the house in the lower-left corner of the map.
4. A person should be drawn in the upper-right corner of the map.
5. Both streets should be labeled.

Week 1 Day 4 (page 18)

1. Exit the room and turn left. Walk down the hall. Turn left at the end of the hall by the stairs to exit.
2. Answers should describe a short, efficient route to get outside.

Week 1 Day 5 (page 19)

Row 1: square in column B and star in column C.
Row 2: circle in column D.
Row 3: Triangle in column A and heart in column C.
Row 4: Diamond in column B.

Week 2 Day 1 (page 20)

1. Africa and Asia
2. Arctic Ocean
3. Responses should include two of the following: Indian Ocean, Pacific Ocean, and Arctic Ocean.
4. Check that all seven continents are shaded in different colors.
5. Check that legends match the colors used to shade corresponding continents.

Week 2 Day 2 (page 21)

1. 650 ft. (200 m)
2. 0 ft. (0 m)
3. 1,650 ft. (500 m)
4. The coasts of Africa have an elevation of 0 ft. (0 m).

Week 2 Day 3 (page 22)

Starting in the Northern Territory and moving clockwise: Northern Territory, Queensland, New South Wales, Tasmania (island), Victoria, South Australia, and Western Australia.

Week 2 Day 5 (page 24)

1. 0°
2. Asia
3. South America
4. Antarctica
5. Europe, Africa, and Antarctica

Week 3 Day 1 (page 25)

1. Answers may include Cabo San Lucus, Puerto Vallarta, or Acapulco.
2. Answers may include Cancún or Veracruz.
3. Mexico City
4. United States, Guatemala, and Belize
5. Acapulco

Week 3 Day 2 (page 26)

Starting at the top and moving clockwise: United States, Gulf of Mexico, Belize, Guatemala, Pacific Ocean

Week 3 Day 3 (page 27)

1. A landfill closed in 2011.
2. They let people trade their recyclable materials for food.
3. The government paid the farmers to give people fruits and vegetables and people were able to trade their trash for food.

Week 3 Day 4 (page 28)

1. United States
2. India
3. United States, Australia, Finland, Turkey, Japan, Mexico, China/Brazil, India
4. Answers may include that people all over the world contribute to the amount of trash, and recycling will decrease the amount that ends up in landfills.

ANSWER KEY *(cont.)*

Week 4 Day 1 (page 30)

1. St. Joseph, MO, and Sacramento, CA
2. California, Nevada, Utah, Wyoming, Nebraska, Colorado, Kansas, and Missouri
3. Answers may include that it made their trips more dangerous.

Week 4 Day 2 (page 31)

From left to right: green, purple, brown, pink, yellow, blue, orange, and red.

Week 4 Day 3 (page 32)

1. 10 days
2. Horses ran 10 to 15 miles before a break. Riders rode 100 miles before a break.
3. As the telegraph was used more, the Pony Express was not needed. Telegraphs were much faster and easier.
4. Answers may include that it was a step forward in faster communication.

Week 4 Day 4 (page 33)

1. Answers may include that the bags hold mail and supplies the riders might need.
2. Answers may include that it was created to remember and honor the Pony Express.
3. Answers may include that the horses had to run very fast or that it is a part of history that people want to remember.

Week 5 Day 1 (page 35)

Answers should include that most Democratic states are in the West or Northeast and most Republican states are in the South and Midwest.

Week 5 Day 2 (page 36)

1. California
2. Wyoming
3. Iowa
4. Alabama
5. Virginia
6. Tennessee

Week 5 Day 3 (page 37)

1. The Electoral College is the group of electors who actually vote for president.
2. States with large populations have more electoral votes. States with smaller populations have fewer of electoral votes.
3. Example: The Electoral College is not fair because the person who receives the most votes should become president.

Week 5 Day 4 (page 38)

1. Answers should include three of the following: Iowa, Wisconsin, Michigan, Ohio, Pennsylvania, and Florida.
2. Answers may include that more people voted for a different party.

Week 6 Day 1 (page 40)

1. Nunavut, Manitoba, Ontario, and Quebec
2. Nunavut is likely colder because it is farther north than Ontario.
3. Alberta, and Saskatchewan
4. Nunavut

Week 6 Day 3 (page 42)

1. Tectonic plates collided, causing the land to buckle and fold.
2. Many of the peaks are shorter and more rounded.
3. The water is important to the wildlife living there.

Week 6 Day 4 (page 43)

1. Denali
2. Mt. Mitchell
3. Mt. Elbert and Mt. Whitney
4. Answers should include that the mountain has been eroded over time.
5. Answers may include that Denali is a young mountain because of its height.

Week 7 Day 1 (page 45)

1. Texas, New Mexico, Oklahoma, Kansas, Colorado, Nebraska, South Dakota, and Iowa.
2. Answers may include that this region experiences many tornadoes.
3. Answers may include that it is in the center of the country or in the Midwest.

ANSWER KEY *(cont.)*

Week 7 Day 3 (page 47)

1. They are ruled on the Enhanced Fujita scale from an EF-0 to an EF-5.
2. Example: *I think the tornado was an EF-3. There is a lot of damage, so it's more than an EF-0. But it isn't completely destroyed, so it's less than an EF-5.*
3. about 1,000 tornadoes

Week 7 Day 4 (page 48)

1. cold, dry air; warm, dry air; warm, moist air
2. Answers may include that the weather is colder farther north.
3. Answers should include that Tornado Alley is where these three types of air meet.

Week 8 Day 1 (page 50)

1. Answers may include fish or crops since it is near a river.
2. Mumbai
3. New Delhi is in the northern central part of the country.

Week 8 Day 3 (page 52)

1. In the past, people often traded to get what they needed. Now, they buy it with money.
2. All of the markets are crowded, sell a variety of items, and are made of individual shops.
3. Example: *I would most like to visit the Palika Bazaar because I think an underground market would be interesting to see.*

Week 8 Day 4 (page 53)

1. Answers may include that both markets are along the sides of a street, both have people walking around, and both are lined with shops.
2. Answers may include that the market today is much more crowded, the shops have moved into the street, and the shops are now covered with canopies.
3. Answers may include that more people are living in the area.

Week 9 Day 1 (page 55)

1. Answers should include two of the following: Mediterranean Sea, Jordan River, Sea of Galilee, or Dead Sea.
2. Most cities are on the coast of the Mediterranean Sea. Reasons may include that there are many resources or that it is easy to trade along the coast.
3. Answers should include two of the following: Egypt, Jordan, Syria, and Lebanon.

Week 9 Day 2 (page 56)

Starting at the top-center of the map and moving clockwise, the countries should be labeled: Lebanon, Syria, Jordan, and Egypt.

Week 9 Day 3 (page 57)

1. Jewish people visit the wall and mourn the temple being destroyed.
2. Muslim people believe that this is where Muhammad was standing when he rose to heaven.
3. It is home to important Jewish and Muslim sites.
4. Answers should be supported with details from the text.

Week 9 Day 4 (page 58)

1. Judaism
2. Answers may include Christians, Buddhists, Hindus, etc.
3. Answers may include that the groups might clash when they have different ideas and values or when deciding who has rights to the region.
4. Answers may include that Israel's diversity helps people see things from many perspectives.

Week 10 Day 1 (page 60)

1. It is in the center of the map.
2. Routes should be as direct as possible.
3. Tokyo Disneyland and Bonsai Park; they are labeled with arrows pointing off the map.
4. Example: *I would most like to visit the Imperial Palace Gardens because I love spending time outside.*

ANSWER KEY *(cont.)*

Week 10 Day 3 (page 62)

1. The walls are made of panels that can move around.
2. Tatami mats cover the floors and are made of soft, woven rice straws.
3. Shoji panels are translucent panels that let light in.
4. Answers may include that people like the traditional styles, they still find them convenient, or people want to honor their heritage.

Week 10 Day 4 (page 63)

1. between 1945 and 1955
2. Answers should explain that although there was a drop in the 1940s, the population has steadily grown.
3. Answers should be slightly higher than the 2015 population (13.49 million).
4. Answers may include that more houses and apartments are needed to house more and more people.

Week 11 Day 1 (page 65)

1. The highest populations are in eastern China.
2. Yellow Sea, East China Sea, and South China Sea.
3. Answers should explain that it is likely in eastern China since it has a higher population.

Week 11 Day 2 (page 66)

Starting at Beijing (marked with a star) and moving clockwise, the cities should be labeled: Beijing, Shanghai, Wuhan (west of Shanghai), Guangzhou (just inland from Hong Kong), Hong Kong, Lijiang, and Lhasa.

Week 11 Day 3 (page 67)

1. The population was growing too quickly, and leaders were afraid the country could not provide for so many people.
2. Answers should include that families were given better jobs and more money for following the policy and were forced to pay fines for not following it.
3. Some daughters were given to orphanages. Sons were preferred so they could carry on the family name.

Week 11 Day 4 (page 68)

1. Answers may include that eastern China is closer to the ocean, the soil or weather may be better, or there are more opportunities in large cities.
2. Answers may include that eastern China is overcrowded, while western China may not have all as many resources.
3. Answers may include that eastern China has such a huge population, and the government wanted to slow the population growth.

Week 12 Day 1 (page 70)

1. Indonesia and the Philippines
2. Myanmar
3. Laos
4. Thailand

Week 12 Day 2 (page 71)

Myanmar, Laos, Thailand, Cambodia, and Vietnam should be shaded. The Philippines and Indonesia should have stripes. Malaysia should be circled.

Week 12 Day 3 (page 72)

1. There is a wet season and a dry season.
2. Answers may include that the plants in the evergreen forest are closer together, and trees in the deciduous forest have long trunks and leaves at the top.
3. Answers may include that people near a tropical evergreen forest need to prepare for rain all year round, while people near a tropical deciduous forest have to prepare for a rainy season and a dry season.

Week 12 Day 4 (page 73)

1. The sugar cane looks like tall grasses or bushes with long leaves. The rice is short and in bunches.
2. The sugar cane is grown in rows on land. The rice is grown in water on tiered paddies.

Week 13 Day 1 (page 75)

1. All of the cities are on the coast.
2. Canberra is the national capital because it has the large star.
3. seven states and territories

Week 13 Day 2 (page 76)

5. Coral Sea, Tasman Sea, Great Australian Bight, Indian Ocean, Timor Sea, Arafura Sea, and Gulf of Carpentaria

ANSWER KEY *(cont.)*

Week 13 Day 3 (page 77)

1. They traveled on horses.
2. The railways were not the same size. Trains could not travel on all of the tracks, only the size they were built for.
3. The first railway connected three miles between Melbourne and Port Melbourne.

Week 13 Day 4 (page 78)

1. Answers may include that railways made travel and trade easier.
2. Answers may include that a standard-size railway made it easier to travel between states.
3. Answers may include that it was expensive or that they may have had to stop train service during construction.

Week 14 Day 1 (page 80)

1. Tennant Creek and Alice Springs
2. Timor Sea
3. Answers may include that it is a landmark.

Week 14 Day 2 (page 81)

1. Drawings should be approximately one-third of an inch tall and four inches long.
2. Drawings should be approximately 2.2 inches long and 1.5 inches wide.

Week 14 Day 3 (page 82)

1. A monolith is a solid, single rock.
2. The Aborigines called it Uluru, and the Europeans call it Ayers Rock.
3. Aborigines are the native people of Australia.
4. Example: *I think people should not be able to climb the rock because it is sacred to the Aborigines.*

Week 14 Day 4 (page 83)

1. Answers may include that it shows people or animals.
2. Answers may include that they want to learn about a different culture.
3. Answers may include that it shows their culture.

Week 15 Day 1 (page 85)

1. more than 150 cm
2. The south central and western regions receives the least amount of rainfall.
3. Answers may include that the coasts receive the most rain.
4. Answers may include that people would want to build a city in an area that receives enough rain.

Week 15 Day 2 (page 86)

Starting at Darwin in the north and moving clockwise, cities should be labeled: Darwin, Brisbane, Sydney, Hobart (on the island of Tasmania), Melbourne, Adelaide, and Perth.

Week 15 Day 3 (page 87)

1. Earth has seasons because it is tilted on its axis.
2. Answers may include that people do summertime activities on Christmas rather than wintertime activities.
3. Answers should include that the United States and Australia are in opposite hemispheres. When Australia is tilted toward the sun and having summer, the United States is tilted away from the sun and having winter.

Week 15 Day 4 (page 88)

1. It falls as rain because it doesn't get cold enough for snow or ice.
2. June, July, August
3. June, July, and August should be circled. December, January, and February should have boxes drawn around them.
4. Answers may include that the seasons are opposite or that it is more or less mild than where they live.
5. Answers should be supported with details from the chart.

Week 16 Day 1 (page 90)

1. Sydney, Newcastle, and Coffs Harbour
2. Tamworth
3. Bourke
4. in the east or along the coast

ANSWER KEY *(cont.)*

Week 16 Day 3 (page 92)

1. There was no room in British jails.
2. They farmed and built roads, buildings, and bridges.
3. Answers may include that they had to start over in a new place without any buildings or crops, or they might have missed friends and family they left behind.

Week 16 Day 4 (page 93)

1. The population for 1800, 1850, 1900, and 1950 would be higher if they were included. And it would account for some of the rapid growth since 1950.
2. 1950 and 2000
3. The UK did not begin sending convicts to Australia until the late 1700s.

Week 17 Day 1 (page 95)

1. Leonora
2. 24°S, 120°E
3. Answers should include three of the following: Mount Gambier, Geelong, Melbourne, Launceston, and Hobart.
4. Answers should be close to 18°S, 146°E

Week 17 Day 2 (page 96)

1. An *A* should be in the Arafura Sea.
2. A *B* should be in West Australia, near Leonora.
3. A *C* should be in Queensland, near Mount Isa.
4. A *D* should be in New South Wales, near Swan Hill.
5. An *E* should be in West Australia, near Derby.
6. An *F* should be in the Northern territory, South of Alice Springs.

Week 17 Day 3 (page 97)

1. Relative location tells where a place is compared to another place. Exact location tells exactly where a place is on the map. Both help people find locations.
2. latitude: equator; longitude: prime meridian

Week 17 Day 4 (page 98)

1. Example: city: Cairns; relative location: northwest of Townsville; absolute location: 16°S, 145°E
2. Answers may include the following: relative location is helpful when latitude and longitude are not available: it can be easier to follow when driving: absolute location is helpful when you need to pinpoint a place's exact location.

Week 18 Day 1 (page 100)

1. Caracas
2. Venezuela is on the northern part of South America because it is shown at the top of the continent.
3. nearest: Calabozo; farthest: Santa Elena de Uairén

Week 18 Day 3 (page 102)

1. gasoline, to make electricity, asphalt roads, and plastic
2. the Orinoco Belt
3. Answers should include that the country makes money off oil, and it provides jobs for many people.

Week 18 Day 4 (page 103)

1. Answers should include that coffee production has remained mostly stable or that it was trending down in 1994, then it started an upward trend.
2. Answers may include that if the weather was too dry or too wet, the plants might have produced fewer beans.
3. Answers may include that there was a large jump in the number of bags produced.

Week 19 Day 1 (page 105)

1. 300–1,000 m
2. Cusco
3. Answers may include that it is harder to reach those heights, or the weather is colder.

Week 19 Day 2 (page 106)

Maps should have a desert strip on the west coast, rainforest in the east, and mountains in the center. Oceans should be colored blue.

ANSWER KEY *(cont.)*

Week 19 Day 3 (page 107)

1. The desert is on the west coast.
2. They provide grass to eat.
3. Answers may include that people in the desert have to conserve water, in the mountains, they have to wear warmer clothes, and in the rainforest, they may need to learn about the different plants.

Week 19 Day 4 (page 108)

1. It would be better to raise animals because there is more land available for pasture.
2. forests
3. Answers may include that the land might be used for roads or cities.

Week 20 Day 1 (page 110)

1. Atlantic Ocean
2. west to east
3. Brazil
4. Answers should inclue two of the following: Negro River, Madeira River, Maranon River, and Xingu River.

Week 20 Day 3 (page 112)

1. GPS and satellites help measure the Amazon River.
2. It begins in the Andes Mountains and ends in the Atlantic Ocean.
3. They might be able to use then to cure sicknesses.

Week 20 Day 4 (page 113)

1. Answers may include that rivers were used to water crops, used for drinking and bathing, and they were a mode of transportation.
2. Answers should correctly identify the continent and length of the river listed.
3. The weather is so cold that most of the water is frozen.
4. Onyz River, Murray River, Volga River, Missouri River, Yangtze River, Amazon River, and Nile River

Week 21 Day 1 (page 115)

1. Chile, Bolivia, Paraguay, Brazil, Uruguay
2. 2,400 miles
3. 1,200 miles
4. 600 miles

Week 21 Day 2 (page 116)

1. A triangle should be drawn two inches north of Río Gallegos.
2. A star should be drawn half an inch east of San Juan.
3. A square should be drawn one inch northwest of Santa Fe.
4. A circle should be drawn $1\frac{1}{2}$ inches southwest of Bahía Blanca.

Week 21 Day 3 (page 117)

1. The northern part is closer to the equator and is tropical. The southern part is far from the equator and is cold.
2. The Pampas are a large area of plains in central Argentina.
3. People grow crops, such as corn, grain, and alfalfa and raise animals, such as sheep, cattle, and horses.
4. Answers should be supported by details from the text.

Week 21 Day 4 (page 118)

1. Answers may include that the people's clothing looks old-fashioned, there are no modern modes of transportation, and the housing does not look modern.
2. Pampas (plains)
3. Answers may include that the land is very flat, it looks like a farm, or there are no visible mountains or ocean.

Week 22 Day 1 (page 120)

1. La Paz
2. Colombia and Peru
3. Answers should include three of the following: Guyana, Suriname, French Guiana, Venezuela, Colombia, Peru, Bolivia, Paraguay, Argentina, or Uruguay.
4. Chile

Week 22 Day 2 (page 121)

Beginning in Brazil and moving clockwise: Brazil, Argentina, Chile, Peru, Colombia, and Venezuela.

ANSWER KEY *(cont.)*

Week 22 Day 3 (page 122)

1. by ship
2. Answers may include that roads connect would make traveling and trading between countries easier.
3. Answers may include that the mountains are steep or that they may need to tunnel through rock.

Week 22 Day 4 (page 123)

1. Paraguay has the fewest cars because it has the most people for each car.
2. Answers may include walking, riding bikes, or public transportation.
3. Answers may include that places with fewer cars may be harder to access and need more roads, or places with many cars need more roads to accommodate so many cars.
4. Answers may include that car companies might try to sell more cars in countries that have more cars.

Week 23 Day 1 (page 125)

1. Northern Cape
2. Answers should include two of the following: Eastern Cape, KwaZulu-Natal, North West, and Mpumalanga.
3. Namibia and Botswana
4. along the western coast and in the center of the country

Week 23 Day 3 (page 127)

1. Kimberley, South Africa
2. Answers may include that the person who found it was excited.
3. Answers may include that they are far underground or the equipment could injure them.

Week 23 Day 4 (page 128)

1. Answers may include that mining provides jobs for people and the diamonds can be sold for profit.
2. Answers may include that larger countries have more space to look for diamonds.
3. Russia
4. 51 million carats
5. Answers should be supported by details from the chart.

Week 24 Day 1 (page 130)

1. Sinai
2. Giza is in the northern part of Egypt, along the Nile River. It is just west of Cairo.
3. Libya, Sudan, and Israel

Week 24 Day 2 (page 131)

Beginning at Siwa (far left) and moving clockwise: Siwa, Giza, Mediterranean Sea, Cairo, Luxor, and Red Sea.

Week 24 Day 3 (page 132)

1. They were built over 4,500 years ago.
2. The pyramids were tombs that held treasure and other items Pharaohs would need in the afterlife.
3. Historians believe that ramps were built around them. Then, workers would use rollers and levers to move the stones into place.
4. The pyramids were built for the Pharaohs Khufu, Khafre, and Menkaure.

Week 24 Day 4 (page 133)

1. Answers may include that the pyramid is very tall.
2. Answers may include that the pyramids were built without modern technology or that the pyramids have not crumbled over time.
3. Answers may include that people want to show how important the monument is.

Week 25 Day 1 (page 135)

1. It is in southeastern Africa.
2. Answers should include the smaller map showing where the country is located.
3. Mutare
4. Botswana, Zambia, Mozambique, and South Africa

Week 25 Day 3 (page 137)

1. English, Shona, and Ndebele
2. They did not want the other languages to die out.
3. Answers may include that it helps preserve the languages, but it might make it difficult for many people to communicate.

Week 25 Day 4 (page 138)

1. a person's first language
2. Answers may include that few people speak those languages.
3. Example: *I think they should keep all 16 languages because otherwise some of the languages could die out.*

ANSWER KEY *(cont.)*

Week 26 Week 1 (page 140)

1. Tanzania
2. Tunisia
3. Chad is north central Africa. It is south of Libya, east of Niger, west of Sudan, and north of the Central African Republic.

Week 26 Day 3 (page 142)

1. They wanted more land and resources.
2. It had thousands of separate states. Each had its own language and culture.
3. Answers may include that Europeans took away their freedom.

Week 26 Day 4 (page 143)

1. Answers may include that there are more countries shown in the modern map or that European countries are not shown in the modern map.
2. Answers may include that they have more control over their governments.

Week 27 Day 1 (page 145)

1. It spans across the northern part of Africa.
2. The climate is very dry and hot because it is a desert.

Week 27 Day 2 (page 146)

Algeria, Egypt, Libya, Mauritania, and Niger should be colored red. Chad, Mali, and Sudan should be colored orange. Botswana and South Africa should be colored yellow. Angola, Cameroon, Ethiopia, Tanzania, and Zambia should be colored green.

Week 27 Day 3 (page 147)

1. It was difficult to travel through the large, harsh desert.
2. Both regions share a religion (Islam) and are more developed.
3. Answers may include that the climates and cultures are different. Saharan Africa has a desert climate and developed like in the Middle East. Sub-Saharan Africa is more diverse.

Week 27 Day 4 (page 148)

1. rainforest, desert, and grassland
2. Example: *People in the rainforest would live with a lot of moisture and greenery. In the desert, people would have to conserve water and stay cool. In the grasslands, they may need to watch out for wild animals.*
3. Example: *I would like to visit the rainforest because I think there would be many interesting plants and animals.*

Week 28 Day 1 (page 150)

1. Spain and the Mediterranean Sea
2. Some beaches are being worn away while others are growing.
3. Answers may include that the water could get too close to homes and buildings.

Week 28 Day 2 (page 151)

1. Paris should be circled.
2. Areas with lighter shading should be marked.
3. Areas with darker shading should be marked.
4. Coastal cities should be boxed.
5. Answers may include that everyone should be concerned because the coasts are an important habitat.

Week 28 Day 3 (page 152)

1. They are eroding into the ocean.
2. Advantages may include that they can stop or slow erosion. Disadvantages may include that they are expensive to build and maintain, and they may not work over time.
3. Example: *I would choose to build a seawall because it can protect people and the environment.*

Week 28 Day 4 (page 153)

1. Open beaches erode most easily because the sand washes away.
2. The headlands have rocks and steep cliffs. Open beaches are flat and sandy. Mudflats and saltmarshes have plants growing on the land.
3. Answers may include that people want to protect the open beaches because they are the most popular to visit.

Week 29 Day 1 (page 155)

1. about 550 km
2. about 400 km
3. about 200 km

ANSWER KEY *(cont.)*

Week 29 Day 2 (page 156)

Beginning at Haapsala in the north and moving clockwise: Haapsala, Pärnu, Ukmerge (farthest south), Panevezys, Liepaja, and Ventspilis.

Week 29 Day 3 (page 157)

1. Estonia, Latvia, and Lithuania
2. All three countries were taken over by the U.S.S.R. in 1940 and became independent in 1991.
3. They were ruled by Russia and are very near to Russia.

Week 29 Day 4 (page 158)

1. Estonian in Estonia, Latvian in Latvia, and Lithuanian in Lithuania
2. Answers may include that it is a diverse region or that people are proud of their cultures.
3. Answers may include that it could help people who speak different first languages communicate.
4. Answers may include that most people still speak Estonian, Latvian, and Lithuanian.

Week 30 Day 1 (page 160)

1. Answers should include that they are in northern Europe.
2. Denmark
3. Norway, Finland, North Sea, Baltic Sea, and Gulf of Bothnia

Week 30 Day 2 (page 161)

A tree should be drawn in Sweden, a wind turbine should be drawn in Denmark, and a wave should be drawn in Norway.

Week 30 Day 3 (page 162)

1. wind energy
2. Renewable resources do not run out and can be used again. Non-renewable resources can be used up.
3. Bioenergy turns energy stored in plants into electricity.

Week 30 Day 4 (page 163)

1. Answers may include that they want to help the planet, and it won't run out.
2. Answers may include that other countries may not have as many renewable resources, or that there are many different types of renewable resources for countries to use.
3. Answers may include to be at 100 percent for all kinds of power, not just electricity or to help other countries use more renewable resources.

Week 31 Day 1 (page 165)

1. England, Scotland, Wales, and Northern Ireland
2. the English Channel
3. Scotland

Week 31 Day 3 (page 167)

1. The U.K. could make more goods than they needed because the inventions helped them make things more easily.
2. Answers should include two of the following: weaving fabric, casting iron, and the steam engine.
3. They traded for tea and spices.
4. The British traded for cotton.

Week 31 Day 4 (page 168)

1. The United Kingdom can't produce everything it needs.
2. Cars, medicine, and oil
3. Answers may include that they import and export different types of cars.
4. It is bad for the economy because they are spending more than they are making.

Week 32 Day 1 (page 170)

1. Answers should include four of the following: Corfu, Rhodes, Ermoupoli, Chania, or Heraklion.
2. Aegean Sea
3. Athens
4. Answers may include islands and peninsulas.

Week 32 Day 2 (page 171)

Beginning on the left side of the map and moving clockwise, the countries should be labeled Albania, Macedonia, Bulgaria, and Turkey.

ANSWER KEY *(cont.)*

Week 32 Day 3 (page 172)

1. Letters may include a, s, k, p, and o.
2. They are used to represent ideas or unknown numbers.
3. Answers may include that the language is very old.

Week 32 Day 4 (page 173)

1. Answers may include that you can use the Greek meaning to figure out what parts of the English word mean.
2. Auto, bio, and graph mean "self," "life," and "write," so it is something you write about your own life.
3. Answers may include trio, tricycle, triple, tripod, triceps, trilogy, etc.
4. Answers may include antibiotic, antiseptic, antidote, antisocial, etc.

Week 32 Day 5 (page 174)

Greek: Σ, Γ, Λ, Δ, Υ, Φ, Ξ, Ψ, Θ, Π, and Ω
English: C, D, F, G, J, L, Q, R, S, U, V, W, and Y
both: A, B, E, Z, H, I, K, M, N, O, P, T, and X

Week 33 Day 1 (page 175)

1. Mongolia
2. Answers may include Kazakhstan, Mongolia, and China.
3. Answers may include Canada, the United States, Mexico, or El Salvador.
4. Answers may include Russia, Kazakhstan, Mongolia, Iran, Saudi Arabia, India, Vietnam, the Philippines, Japan, or Indonesia.

Week 33 Day 2 (page 176)

A solid line should be drawn from England to the United States. Dashed lines should be drawn from the United Kingdom, Ireland, Germany, Japan, and China to the United States. Dashed and dotted lines should be drawn from Mexico, China, India, the Philippines, Vietnam and El Salvador to the United States.

Week 33 Day 3 (page 177)

1. A person who moves from one country to another.
2. There was a potato famine, and people did not have enough to eat.
3. People might not feel safe in their own country.

Week 33 Day 4 (page 178)

1. 1980 and 2010
2. The number of immigrants decreased. The numbers usually increase. Reasons may include global conditions or that the United States allowed fewer immigrants to come.
3. Answers may include that the numbers will continue to rise because that is the trend of the chart.

Week 34 Day 1 (page 180)

1. Brazil
2. Answers may include 45°N, 90°E; 30°N, 90°E; 45°N, 120°E; and 30°N, 120°E.
3. Answers should include two of the following: Russia, China, Philippines, Indonesia, and Australia.
4. Answers should include two of the following: United States, Canada, and Russia.
5. Answers may include anything between about 10°S–45°S and 110°E–155°E.

Week 34 Day 2 (page 181)

1. An *A* should be written in the northwestern United States.
2. A *B* should be written in central Russia.
3. A *C* should be written in the South Pacific Ocean.
4. A *D* should be written in southern India.
5. An *E* should be written near Madagascar.

Week 34 Day 3 (page 182)

1. They measured shadows to find the angle of the sun.
2. A perfect clock was needed to time one hour, which equaled 15 degrees.
3. Answers may include that it made navigation safer and more accurate.

Week 34 Day 4 (page 183)

1. They are curved because Earth is a sphere.
2. Answers may include that North and South America are both too wide on the old map.
3. Answers may include that exploration was dangerous since people did not know the true shape of the land.

Week 35 Day 2 (page 186)

Use the map on page 207 to check student answers.

ANSWER KEY *(cont.)*

Week 35 Day 3 (page 187)

Use the map on page 207 to check student answers.

Week 35 Day 4 (page 188)

Use the map on page 207 to check student answers.

Week 35 Day 5 (page 189)

Use the map on page 207 to check student answers.

Week 36 Day 2 (page 191)

1. Hartford
2. Des Moines
3. Topeka
4. Augusta
5. Boston
6. Saint Paul
7. Jefferson City
8. Lincoln
9. Concord
10. Trenton
11. Albany
12. Bismarck
13. Harrisburg
14. Providence
15. Pierre
16. Montpelier

Week 36 Day 3 (page 192)

1. Montgomery
2. Little Rock
3. Tallahassee
4. Atlanta
5. Springfield
6. Indianapolis
7. Frankfort
8. Baton Rouge
9. Lansing
10. Jackson
11. Raleigh
12. Columbus
13. Oklahoma City
14. Columbia
15. Nashville
16. Austin
17. Madison

Week 36 Day 4 (page 193)

1. Juneau
2. Phoenix
3. Sacramento
4. Denver
5. Dover
6. Honolulu
7. Boise
8. Annapolis
9. Helena
10. Carson City
11. Santa Fe
12. Salem
13. Salt Lake City
14. Richmond
15. Olympia
16. Charleston
17. Cheyenne

Week 36 Day 5 (page 194)

Use the map on page 207 to check student answers.

POLITICAL MAP OF THE UNITED STATES

PHYSICAL MAP OF THE UNITED STATES

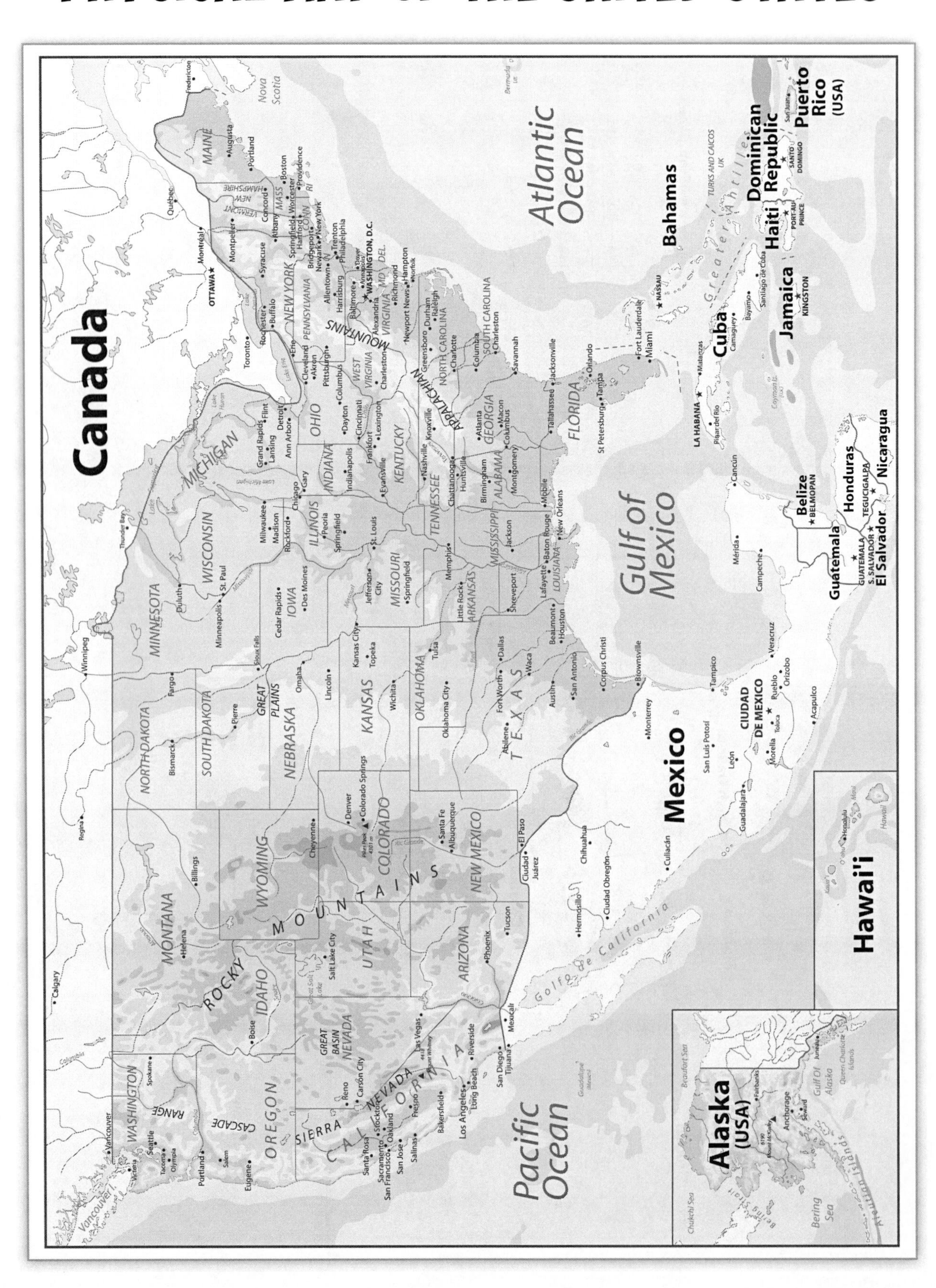

WORLD MAP

Alaska (U.S.A.)
Canada
United States
Greenland
Mexico
Guatemala
El Salvador
Costa Rica
Panama
Honduras
Nicaragua
Cuba
Colombia
Venezuela
Peru
Bolivia
Brazil
Paraguay
Chile
Argentina
Uruguay
Iceland
Ireland
United Kingdom
Norway
Sweden
Finland
Denmark
Germany
Poland
Belarus
France
Spain
Italy
Ukraine
Romania
Bulgaria
Georgia
Turkey
Syria
Iraq
Iran
Russia
Kazakhstan
Uzbekistan
Turkmenistan
Afghanistan
Pakistan
Mongolia
China
Nepal
India
Myanmar
Laos
Thailand
Cambodia
Vietnam
Taiwan
Philippines
Malaysia
Indonesia
Japan
Australia
New Zealand
Morocco
Algeria
Mauritania
Mali
Guinea
Niger
Nigeria
Libya
Chad
Cameroon
Egypt
Sudan
Congo
Saudi Arabia
Yemen
Ethiopia
Somalia
Kenya
Tanzania
Angola
Zambia
Namibia
Bostania
Zimbabwe
South Africa
Mozambique
Madagascar

Name: ________________________ Date: ____________

MAP SKILLS RUBRIC
DAYS 1 AND 2

Directions: Evaluate students' activity sheets from the first two weeks of instruction. Every five weeks after that, complete this rubric for students' Days 1 and 2 activity sheets. Only one rubric is needed per student. Their work over the five weeks can be evaluated together. Evaluate their work in each category by writing a score in each row. Then, add up their scores, and write the total on the line. Students may earn up to 5 points in each row and up to 15 points total.

Skill	5	3	1	Score
Identifying Map Features	Uses map features to correctly interpret maps all or nearly all the time.	Uses map features to correctly interpret maps most of the time.	Does not use map features to correctly interpret maps.	
Using Cardinal Directions	Uses cardinal directions to accurately locate places all or nearly all the time.	Uses cardinal directions to accurately locate places most of the time.	Does not use cardinal directions to accurately locate places.	
Interpreting Maps	Accurately interprets maps to answer questions all or nearly all the time.	Accurately interprets maps to answer questions most of the time.	Does not accurately interpret maps to answer questions.	

Total Points: ____________________

Name: ______________________ Date: ______________

APPLYING INFORMATION AND DATA RUBRIC

DAYS 3 AND 4

Directions: Complete this rubric every five weeks to evaluate students' Day 3 and Day 4 activity sheets. Only one rubric is needed per student. Their work over the five weeks can be evaluated together. Evaluate their work in each category by writing a score in each row. Then, add up their scores, and write the total on the line. Students may earn up to 5 points in each row and up to 15 points total. **Note:** Weeks 1 and 2 are map skills only and will not be evaluated here.

Skill	5	3	1	Score
Interpreting Text	Correctly interprets texts to answer questions all or nearly all the time.	Correctly interprets texts to answer questions most of the time.	Does not correctly interpret texts to answer questions.	
Interpreting Data	Correctly interprets data to answer questions all or nearly all the time.	Correctly interprets data to answer questions most of the time.	Does not correctly interpret data to answer questions.	
Applying Information	Applies new information and data to known information about locations or regions all or nearly all the time.	Applies new information and data to known information about locations or regions most of the time.	Does not apply new information and data to known information about locations or regions.	

Total Points: ______________________

Name: ______________________________ Date: ________________

MAKING CONNECTIONS RUBRIC
DAY 5

Directions: Complete this rubric every five weeks to evaluate students' Day 5 activity sheets. Only one rubric is needed per student. Their work over the five weeks can be evaluated together. Evaluate their work in each category by writing a score in each row. Then, add up their scores, and write the total on the line. Students may earn up to 5 points in each row and up to 15 points total. **Note:** Weeks 1 and 2 are map skills only and will not be evaluated here.

Skill	5	3	1	Score
Comparing One's Community	Makes meaningful comparisons of one's own home or community to others all or nearly all the time.	Makes meaningful comparisons of one's own home or community to others most of the time.	Does not make meaningful comparisons of one's own home or community to others.	
Comparing One's Life	Makes meaningful comparisons of one's daily life to those in other locations or regions all or nearly all the time.	Makes meaningful comparisons of one's daily life to those in other locations or regions most of the time.	Does not make meaningful comparisons of one's daily life to those in other locations or regions.	
Making Connections	Uses information about other locations or regions to make meaningful connections about life there all or nearly all the time.	Uses information about locations or regions to make meaningful connections about life there most of the time.	Does not use information about locations or regions to make meaningful connections about life there.	

Total Points: ______________________________